Climate Cultures in Europe and North America

Bringing together scholarly research by climate experts working in different locations and social science disciplines, this book offers insights into how climate change is socially and culturally constructed.

Whereas existing studies of climate cultural differences are predominantly rooted in a static understanding of culture, cultural globalization theory suggests that new formations emerge dynamically at different social and spatial scales. This volume gathers analyses of climate cultural formations within various spaces and regions in the United States and the European Union. It focuses particularly on the emergence of new social movements and coalitions devoted to fighting climate change on both sides of the Atlantic. Overall, *Climate Cultures in Europe and North America* provides empirical and theoretical findings that contribute to current debates on globalization, conflict and governance, as well as cultural and social change.

This book will be of great interest to students and scholars of climate change, environmental policy and politics, environmental sociology, and cultural studies.

Thorsten Heimann is a sociologist, cultural scientist, and Research Associate at the Research Center for Sustainability, Freie Universitat Berlin, Germany. Since 2020 he has been a strategic advisor for "Green Culture and Sustainability" for the German Minister of State for Culture and the Media in Germany.

Jamie Sommer is an Assistant Professor of Sociology at the University of South Florida.

Margarethe Kusenbach is Professor of Sociology at the University of South Florida.

Gabriela Christmann is a sociologist and Head of the Research Group "Social Innovations in Rural Areas" at the Leibniz Institute for Research on Society and Space (IRS) in Erkner (near Berlin), Germany.

Routledge Advances in Climate Change Research

For more information about this series, please visit: www.routledge.com/
Routledge-Advances-in-Climate-Change-Research/book-series/RACCR

Climate Cultures in Europe and North America

New Formations of Environmental Knowledge and Action

Edited by
**Thorsten Heimann, Jamie Sommer,
Margarethe Kusenbach
and Gabriela Christmann**

LONDON AND NEW YORK

First published 2023
by Routledge
4 Park Square, Milton Park, Abingdon, Oxon OX14 4RN

and by Routledge
605 Third Avenue, New York, NY 10158

Routledge is an imprint of the Taylor & Francis Group, an informa business

British Library Cataloguing-in-Publication Data
A catalogue record for this book is available from the British Library

Library of Congress Cataloging-in-Publication Data
A catalog record has been requested for this book

ISBN: 978-0-367-51314-6 (hbk)
ISBN: 978-1-003-05332-3 (pbk)
ISBN: 978-1-003-30700-6 (ebk)

DOI: 10.4324/9781003307006

Typeset in Times New Roman
by codeMantra

Contents

Figures

Tables

About the contributors

Gabriela Christmann, Professor Dr, is a sociologist. She is head of the research group "Social Innovations in Rural Spaces" at the Leibniz Institute for Research on Society and Space, Erkner (near Berlin), Germany. She is also an adjunct professor at the Technische Universität Berlin, Germany. Her research interests include urban and regional sociology, spatial theories, social innovation, social aspects of climate change, and qualitative methods.

Lawrence Culver, Professor Dr, is an historian. He is an Associate Professor in the Department of History at Utah State University, USA. He received his doctorate in History at the University of California, Los Angeles (UCLA). His research interests include environmental history, climate history, urban history, cultural history, and the histories of the USA and North America.

Dana Giesecke, Dipl.-Soz., MSc, is a sociologist. She is the scientific director of "FUTURZWEI. Stiftung Zukunftsfähigkeit" in Berlin, Germany, since 2011. Previously, she was, among other things, head of the office of the German Sociological Association. Dana studied sociology with a focus on cultural sociology, art history, and communication science at the Technical University of Dresden, Germany.

Michael A. Haedicke, PhD, is a sociologist. He is an associate professor in the Sociology Department at the University of Maine, United States, and a faculty fellow at the Senator George J. Mitchell Center for Sustainability Solutions, also at the University of Maine. His research interests include environmental planning and governance, climate adaptation, equity and resilience in food systems, and food- and environment-oriented social movements.

Thorsten Heimann, Dr, is a cultural scientist and sociologist. He is research associate at the Research Center for Sustainability, Freie Universität Berlin, Germany. In his research, he examines socio-spatial and cultural differences in behavior related to environment and climate change. Since 2020, Thorsten has been a strategic advisor for "Green Culture and

Sustainability" for the German Minister of State for Culture and the Media, Berlin.

Jouni J. K. Jaakkola, MD, DSc, PhD, is a Professor of Public Health at the University of Oulu, Finland, and a Research Professor at the Finnish Meteorological Institute. He is the director of the Center of Environmental and Respiratory Health Research (CERH) at the University of Oulu. He has a deep and long-term interest in global health issues. In the early 1990s, he started an international academic career, working in Norway, Russia, the United States, Sweden, and the United Kingdom. In 2008, he established the Center for Environmental and Respiratory Health Research (CERH) at the University of Oulu. In 2014, CERH was designated as a WHO Collaborating Centre in Global Change, Environment and Public Health. His professional mission is to conduct research on topics which help to solve emerging global public health problems.

Suvi Juntunen is working at the Center of Environmental and Respiratory Health Research (CERH), University of Oulu, Finland. She is a researcher in the projects Arahat, investigating Saami climate change adaptation practices, and SosClim, studying climate justice from the perspective of the Saami. Her educational background is cultural geography, and her special research interests include cultural effects of climate change, traditional knowledge, and biocultural diversity. Suvi has worked as a project leader in the Saami organization Sámi Árvvut and as an adviser in the Finnish Saami Parliament focusing on international affairs, traditional knowledge, and biodiversity issues.

Sarah Kessler, M phil, is a social scientist and doctoral candidate at the Department of Geography, Ludwig-Maximilians-Universität (LMU) Munich, Germany. In her PhD project, she investigates societal receptions of climate change and climate-cultural diversity as well as issues of responsibility, efficacy, and knowledge regarding climate protection. Methodologically, she is interested in advancing the application of qualitative research tools such as focus groups and media analyses to the study of climate change and climate action. The role of citizens in climate science also receives attention in her growing body of work.

Margarethe Kusenbach, PhD, is a professor in the Department of Sociology at the University of South Florida, USA. Her publications and research interests fall into the areas of cities and communities, emotions and identities, environment and disasters, as well as qualitative research methods. Her current research focuses on public art, neighbor culture, and sustainable urban development in American and European cities.

Doris McGonagill, PhD, holds a position as Associate Professor of German at Utah State University, USA. She received her PhD in Germanic

Languages and Literatures from Harvard University. Her interdisciplinary research focuses on the intersection between German literature, the visual arts, aesthetic theory, and memory theory in the 20th and 21st centuries. More recently, she has ventured into the fields of German environmental literature and ecocriticism.

Pilar Morales-Giner, MA, is a sociologist. She is a PhD candidate in the Department of Sociology, Criminology and Law at the University of Florida in Gainesville (USA). Pilar is also a Graduate Assistant at the Office of Research and Assessment in the Division of Student at the University of Florida. Her research interests include environmental sociology, sense of place, responses to climate change, environmental conservation, and environmental governance.

Klemetti Näkkäläjärvi, PhD, is a Saami researcher from Enontekiö, North Finland. He is a senior researcher in the SosClim project at the Center of Environmental and Respiratory Health Research (CERH) at the University of Oulu, Finland, studying climate justice from a Saami perspective. His educational background is cultural anthropology, Saami culture, and language. Klemetti's research expertise includes climate change, climate ethnography, traditional knowledge, biocultural diversity, Saami reindeer herding culture, and Saami language. He has worked on various research projects and was co-project leader in the International Centre for Reindeer Husbandry (ICR) in 2020. He was also the full-time president of the Finnish Saami Parliament between 2008 and 2015, and he is a trusted advisor when it comes to issues such as the Saami language, culture, traditional knowledge, and biodiversity.

Henrike Rau, Professor Dr, is a social scientist and Professor of Social Geography and Sustainability Research at the Department of Geography, Ludwig-Maximilians-Universität (LMU) Munich, Germany. She has made internationally recognized contributions to the conceptual and methodological advancement of social scientific and interdisciplinary sustainability research on topics such as domestic energy use, food consumption, and mobility practices. She is currently leading mixed methods research on public perceptions of climate change and related questions of climate-cultural diversity as part of BAYSICS, an interdisciplinary project that focuses on citizen engagement in climate change research (www.baysics.de).

Miriam Schad, Dr, is a sociologist and a researcher at the Faculty of Social Science, Technical University Dortmund. Before, she worked at the Institute for Advanced Study in the Humanities (KWI) Essen and studied sociology and economics at the Philipps University of Marburg. In her research, she focuses on social inequality (status maintenance in the middle class and precarity), environmental sociology (in relation to inequality and populism), and research on transformational conflicts.

Hans-Georg Soeffner, Professor Dr, is a sociologist. He is a senior professor at the Institute for Political Science and Sociology at the Rhenish Frie- drich Wilhelm University of Bonn (Germany). He is also a board member and permanent fellow at the Institute for Advanced Study in the Human- ities in Essen (Germany). His research focuses on sociological theory, everyday cultures, sociology of knowledge, culture, media, and religion, as well as methodology and methods of hermeneutics in the sociology of knowledge.

Bernd Sommer, Dr habil, is a sociologist. He is head of the research divi- sion "Climate, Culture and Sustainability" at the Norbert Elias Center, Europa-Universität Flensburg, Germany. In his research, he has special- ized in the social dimensions of climate change and the transformation of modern societies toward a more sustainable relationship with nature.

Jamie Sommer, PhD, is a sociologist. She is an Assistant Professor in the De- partment of Sociology at the University of South Florida, Tampa, USA. Her research uses mixed methods to examine how institutional factors impact global inequality in environment and development outcomes. Broadly, she is interested in environmental sociology, global inequality, global political economy, state capacity, and resilience.

Julia Teebken, Dr, is a political and social scientist. She is currently a post-doctoral researcher in the Peking-Princeton Postdoctoral Program at the Department of Environmental Management at Peking University and Princeton University. As a comparativist, her research focuses on policy-making and social practices related to climate change (adapta- tion) and inequality across different political systems.

Harald Welzer, Professor Dr, is a social psychologist. He is head of the Norbert-Elias-Center for Transformation Design at the Europa- Universität Flensburg and head of "FUTURZWEI. Stiftung Zukunfts- fähigkeit" in Berlin, Germany. His research focuses on social impacts of climate change, mass violence, and potentials for socio-ecological trans- formation. His books have been translated into 22 languages.

David Zeller, PhD, is a sociologist. He is an Assistant Teaching Professor at the University of Tampa in Tampa, FL, USA. His research interests include social psychology, social movements, environmental sociology, urban sociology, deviance, theory, and both qualitative and quantitative methods.

Part I
Introduction

Examining Climate Cultures in Europe and North America

Jamie Sommer, Thorsten Heimann,
Margarethe Kusenbach and Gabriela Christmann

This edited volume assembles theoretical papers and empirical studies on climate cultures in Europe and North America. Serving as the key concept for the book, we define "climate cultures" as collectively shared *perceptions* regarding climate change, combined with related social *practices* (Heimann 2019, 19). As discussed below, previous research on the topic demonstrates that there are significant variations in climate change perceptions and related practices at varying social levels as well as geographic scales. Values, views, and understandings of climate change are not necessarily tied to science-based knowledge and solutions, nor do they seamlessly translate into individual or collective action.

Several key edited volumes published over the past dozen years illustrate that social science disciplines have finally begun to capture the essence of the challenge climate change poses to our world. For example, in Szerszynski and Urry's (2010) special issue of the journal *Theory, Culture & Society*, contributors examine how climate science has been produced, institutionalized, mobilized, and contested in the past, as well as exploring the relationships between climate change, politics, global inequities, and financial markets. Focusing more on material aspects, contributors to a volume in the book series *Research in Economic Anthropology*, edited by Wood (2015), explore how people around the world have culturally adapted, or rather failed to adapt, to shifts in economic conditions caused by climate change. These, as well as many other publications, including the previous volumes in this book series, show that the social sciences offer important contributions to uncovering political, social, and economic aspects and outcomes of climate change.

In this book, we aim to build on these and other significant works while zooming in on the particular question of how climate cultures, as defined above, impact climate change perception, mitigation, and adaptation. Besides two background papers, our volume offers a series of empirical case studies of contemporary climate cultures operating at diverse levels of social organization and in various geographies, in Europe and North America. These two world regions are of special interest to climate change researchers, given their substantial political and economic power, their overuse and waste of natural resources, and their often lacking or conflicted commitment

DOI: 10.4324/9781003307006-2

to climate action, meaning activities that foster climate change mitigation and adaptation in local and global spheres.

Climate cultural differences deeply divide Europeans and North Americans, both *within* and *along* local, regional, and national borders. When comparing today's array of climate perceptions and actions in both Europe and North America, it becomes obvious that new forms of climate cultures have developed over the past two decades and are still emerging and changing at a fast pace. Complex and dynamic differences in climate cultures are observable within varying social and spatial contexts. These contexts often transcend traditional units of analysis, such as local, regional, or national territories, and encompass new social spaces, including professional, subcultural, virtual, or otherwise mediated spheres. To illuminate the richness and variety of contemporary climate cultures, we have brought together a diverse group of social science scholars to address the questions of (1) how, why, and where these cultures emerge, (2) whom and what exactly they entail, (3) how disagreement, rivalry, and conflicts develop within and between different cultures, and (4) what implications climate cultural dynamics hold for the various parties involved, including the natural environment itself.

Climate cultures in previous research

Past work on climate cultures has illustrated the promise and capability of this concept in describing social knowledge and action surrounding climate change. Overall, the literature on climate cultures is distributed over several disciplines and fields, including, among others, anthropology, sociology, political science, and geography, making it challenging to summarize.

One line of inquiry into climate cultures employs a framework called "culture as relational space". Heimann (2019) and others argue that cultures can be understood and examined as forms of "shared knowledge" which also includes practices. Heimann defines cultures based on areas of concern, for example, he distinguishes between "food cultures", "learning cultures", and "climate cultures" (Heimann 2019, 30). Climate cultures, more specifically, are formed by shared perceptions regarding climate change ("vulnerability constructions") as well as shared corresponding practices ("resilience constructions"). Building on this definition, in his (translated) book, Heimann (2019) examines diverse climate cultures in Europe by examining forms and contents of shared knowledge across regional and national borders, and by asking which aspects of climate change skepticism or, respectively, climate change recognition, are being shared by whom and why. Several other researchers have built on, and further developed, Heimann's ideas related to climate cultures, exploring how various cultural clusters have evolved, for instance, through the analysis of historical discourse (Heimann et al. 2021; Bembnista and Heimann 2020; Christmann and Heimann 2017; Hulme 2016; Christmann et al. 2014; Christmann and Ibert 2012), through investigating shared forms of background knowledge, or through studying relational networks and contexts (Heimann 2019; Heimann and Mallick 2016).

Past edited volumes, most notably the one edited by Barnes and Dove (2015), have examined other important aspects of climate cultures, for instance, looking at how climate change was historicized, imagined, and understood within governments and scientific communities. Other research on climate cultures reveals actors' beliefs, values, and identities and might thus inform efforts of addressing cultural change (Heimann 2019; Heimann and Mallick 2016; Hulme 2016; Hoffmann 2015; Dunlap and McCright 2010; Kollmuss and Agyeman 2002). Social researchers have investigated the connections between various forms of shared knowledge and environmental action since at least the 1970s, when environmental sociology and related fields began to emerge (Douglas and Wildavsky 1982; Dunlap and van Liere 1977; Heberlein and Shelby 1977; see Heimann 2019, 84 for an overview).

In recent scholarly discourse on climate change, we observe an increasing emphasis on practices. Some authors use the phrase "everyday climate cultures" in connection with social practice theory while examining ways of recognizing and reducing resource use in everyday life (Evans et al. 2020; Goodman et al. 2020; Smit 2015; Liu et al. 2013; Shove et al. 2012). Most previous research into climate cultures examines shared knowledge and actions that form around experiencing climate impacts, for instance, sea-level rise in Oceania (Crook and Rudiak-Gould 2018), the implications of ideological framings of climate knowledge in British media (Carvalho 2007), or local discursive practices in Brazil, South Africa, and China (Nash et al. 2020). More recent critical work has focused on the significance of practitioners' social location and positionality in climate-related knowledge and action (Ford and Norgaard 2020). In sum, the existing previous literature on climate cultures indicates that there are multiple knowledge systems concerning views of climate change and the environment, many of which can be multi-directionally contradictory to human and environmental well-being.

Contributions of the book

Building on these and other previous works, our volume emphasizes the unequal and stratified character of climate cultures. Those who share power and status across spatial contexts are also likely to share similar values, perceptions, and actions concerning climate change. Consistent climate cultures can be found among global elites, at shared levels of income, age, education, or occupation, besides other structural variables. A group may be concerned about the climate and may want to protect the natural environment; however, there can be a considerable disconnect between what is valued and what is actually done about it. This mismatch may be due to how a problem is imagined or defined, what exactly is included in an issue's understanding, and what is not, or how other values or material constraints (internal or external) may limit or complicate one's options. Climate cultures also become particularly important when we seek to understand how these values vary from one place or social sphere to the other and how our understandings intersect with our, more or less implicit, value systems.

Our volume seeks to understand climate cultures within different social locations in all their complexity, rather than relying on fixed and unreflected geographical units. This flexible focus may help to reveal global hierarchies of knowledge, and how such stratified clusters of knowledge relate to climate action and inaction – in support of previous research on globalization which suggests that these patterns exist. We find value in the concept of climate cultures because it helps us overcome false dualisms. For instance, nuanced analyses of climate cultures reveal that there are more than two groups (those "for" and "against" the environment), that there are overlaps between various groups, as well as contradictions within groups, that escape easy polarization and contrast.

We believe that studying climate cultures and the diverse and patterned ways in which they manifest, offers opportunities for uncovering convergence and divergence in climate change perceptions and (in)actions – within, between, and across spaces, places, and social locations. Detecting shared patterns of knowledge and action across seemingly different contexts, as well as finding nuances and variations within shared realms, improves our understanding of how cultures are constructed, maintained, and legitimized in relation to historical, global, and local power dynamics. Examining climate cultures also offers important insights into similarities and differences across the varying domains of climate change perceptions, values, and actions. Lastly, the concept of climate cultures encourages us to reevaluate what we believe is already known about climate change, and about people's perceptions of opinions. Moreover, the concept may help us to critically examine our own values, beliefs, knowledge, and practices in relation to climate change and to detect potential contradictions and ironies between these domains. If we are honest with ourselves, we find that values, knowledge, and actions on complex social issues are often incongruent; it is crucial to understand how and why these fractures emerge, in our own and others' climate cultures.

Volume overview

The origins of the volume go back to 2018, when Margarethe Kusenbach, Gabriela Christmann, and Thorsten Heimann received funding for a working group from the German organization SDAW (Foundation for German American Academic Relations).[1] This grant supported sessions, workshops, and planning meetings at a "Brainstorming Conference on Transatlantic Research" held in Toronto in March 2019. Jamie Sommer was invited to join the editorial team in the fall of 2019. Unfortunately, throughout 2020 and 2021, the writing, peer-reviewing, revising, and finalizing of chapters were repeatedly delayed due to the COVID epidemic. We are very grateful for the support we received for this project from SDAW, our institutions, colleagues, friends, and family, as well as the excellent staff at Routledge. Above all, we thank our fantastic authors and reviewers for their persistence and patience.

It should not go unmentioned that central theoretical ideas included in the volume were developed as part of the research project "Socio-Cultural Constructions of Vulnerability and Resilience", which was funded by the German Research Foundation (under project number 277230079). The project was conducted by Gabriela Christmann and Thorsten Heimann from 2015 to 2019 at the Leibniz Institute for Research on Society and Space in Erkner (near Berlin), Germany.

Following the introduction, the volume assembles nine chapters (seven original papers and two translations) written by a total of sixteen authors working in academia as well as in applied fields. A majority of our authors, ten out of sixteen, live in Europe which, however, does not mean that their research is similarly bound. Together, our authors represent the disciplines of history, languages, anthropology, public health, geography, political science, and sociology, with the majority, just like the four editors, working in the latter discipline. Authors include a mix of doctoral students, recent PhDs, junior and mid-career scholars, as well as senior and retired colleagues. Another aspect of diversity lies in the variety of (predominantly qualitative) research methods employed by chapter authors, including theoretical reflection, historical comparison, ethnography, interviews (with community members or experts), as well as social media and mass media content analysis.

In Part II of the volume, we have paired two chapters that offer broader perspectives on transatlantic climate cultures. Here, we find a visionary essay on the urgency of climate research in the social and cultural sciences and a comparative historical analysis of climate thinking in Europe and the United States. These papers form a contextual background for the following empirical studies.

The insightful opening essay, written by sociologists Harald Welzer, Hans-Georg Seffner, and Dana Giesecke, was first published in 2010 as the introduction to a German-language volume on climate cultures, edited by the same authors. In it, they deliver a stern wake-up call to readers, in general, and to colleagues in the social and cultural sciences, in particular, regarding the unprecedented impacts climate change is destined to deliver on human lives and societies. While reminding us of already observable, massive natural and social consequences of global warming, as well as warning of those yet to come, the authors ponder what these transformations imply about the nature and future of humanity. Welzer, Seffner, and Giesecke believe that the social and cultural disciplines, and the humanities, can, and must, make essential contributions to the public debate and practical treatment of the problem. They call upon these scholars to abandon imaginary theoretical battles and join the frontlines of those fighting for human survival in the real world before it is too late.

The second background chapter was written by historian Lawrence Culver and German language and culture scholar Doris McGonagill. In rich brushstrokes, the paper paints a vivid image of the movement of climate knowledge and discourse back and forth across the Atlantic, from the time of Enlightenment to the present. The authors argue that the intense mutual exchange of cultural, scientific, and political ideas between the emerging United States and, in particular, Germany created the foundation for today's diversity of climate cultures in North America and Western Europe. Throughout history, political and economic regimes in both locations – including colonialism, slavery, and "manifest destiny" in the United States, and recurring periods of progressivism and fascism/authoritarianism in Germany – managed to instrumentalize climate knowledge and beliefs in significant ways that can still be felt and observed today. Culver and McGonagill highlight important events and turning points in the histories of both countries, such as the "Year without Summer" (1816) or the fallout from the Chernobyl nuclear disaster (1986), that has led to connected, yet diverse and partially contradictory beliefs on climate, thereby deeply anchoring the following social science chapters in the cultural and political histories of both regions.

Part III of our volume includes four case studies on various aspects of climate cultures conducted in a range of locations, and at different scales, in Europe: two in Germany, one in Spain, and one in Finland. Within the section, chapters are ordered by scope, moving from studies at the national and regional level to smaller, community-level studies within specific locations. In these chapters, place is treated as an integral and complex aspect of social life and organization in the wake of climate change, not merely as a container.

In the section's first chapter, geographer Sarah Kessler and political scientist Henrike Rau depart from the assumption that mass media coverage is a major source of societal knowledge about climate change. At the same time, they note that climate-related content and interpretations offered in the mass media are understudied by social scientists. The authors address this gap by examining different media formats and representations of climate change in Germany around the 2019 European Parliament elections, focusing on how competing interpretations of climate change enter public discourse and how diverging climate cultures (within a nation) emerge as a result, based on context-specific and group-specific differences in the perception and handling of climate change. Empirically, the study relies on a qualitative content analysis of four interlinked data sources, assembled from TV (political talk shows), social media (Facebook and Twitter), print and online news articles (from political weekly magazines), and YouTube videos (from two well-known influencers). The authors discover that, depending on the media format, there are clear differences in how climate change-related challenges are presented, as well as finding differences in the nature and urgency of proposed countermeasures. In sum, the analysis reveals a significant gap between *elite* climate cultures and a variety of climate cultures "*from below*" in the general public, pointing to a considerable diversity of mediated climate cultures within Germany.

In the second chapter, sociologists Bernd Sommer and Miriam Schad examine how civil society actors become what they call "urban change agents" and induce transformations concerning climate change-related ideas and practices. The chapter is based on twelve qualitative interviews with participants in two communities of Cologne (Germany) on their social and habitual characteristics. The authors' analysis, rooted in Bourdieu's theory of practice, shows that actors who are well integrated into socio-spatial networks, locally as well as beyond, have the power to significantly influence local discourse on sustainability and climate protection. The findings show that both social and economic capital are key factors in the commitment and effectiveness of local change agents. Overall, the chapter by Sommer and Schad indicates that climate cultures emerge and proliferate through the relational social networks of agents and that territorial, in this case urban, contexts are somewhat less relevant.

Third, sociologist Pilar Morales-Giner analyzes how rural inhabitants of the province of Granada, Spain, experience and express alienation from their land, each other, and themselves, as consequences of globalization and climate change, making it difficult for them to develop viable adaptations. The author's twenty-four interviews with residents reveal how they had to change their farming model, from traditional irrigation and livestock-based practices to (wasteful) mechanization, in order to live off their crops. This agrarian industrialization has led to an abandonment of sustainable farming practices (or farms altogether), reduced social cohesion, and caused an increase in rural unemployment and population loss. Overall, Moralez-Giner's paper shows how climate cultures are shaped by global political processes, leading to a disconnect between knowledge and feelings about climate change on the one hand and actual adaptation and mitigation practices on the other.

In the section's last chapter, Klemetti Näkkäläjärvi, Suni Juntunen, and Jouvi Jaakkola, a team of authors including anthropologists and public health researchers, take readers to Northern Europe, an area where cultural patterns of life have been impacted by European colonialism for centuries. The authors offer a rare analytic view on indigenous climate cultures, by describing climate change perception and adaptation practices among thirteen Reindeer Saami communities in Sápmi, Northern Finland. The authors examine how cultural developments are directly linked to environmental change, demonstrating that communities can transform rapidly due to alterations in and of the physical landscape itself. Through ethnographic fieldwork conducted in 2015–2019, the authors develop a typology of adaptation models and retrace how Reindeer-related practices have transformed due to changing climate conditions in the second half of the 20th century. They demonstrate that traditional environmental knowledge and practices become more and more obsolete in a rapidly changing environment. A central concern for the future is whether and how the cultural heritage and richness of Samish knowledge and practice will be able to survive in, and adapt to, a changing climate.

Lastly, Part IV of the book gathers three case studies on climate cultures in North America, likewise on various aspects of the phenomenon, and unfolding in different locations and social spheres. As in Part III, chapters are ordered by scope, moving from studies at the national scale to community-level studies within specific locations.

The section opens with sociologist David Zeller's analysis of disagreement within the United States environmental movement over geoengineering proposals. Zeller shows how so-called frame resonance disputes illuminate the development and maintenance of socially shared knowledge constructions about climate change. Drawing on geoengineering-related internet and media discourse of sixteen environmental movement organizations (EMOs), Zeller illustrates that there are two main points of contention relating to frame resonance: (1) whether geoengineering should be discussed at all as a solution, and (2) how geoengineering should be defined and possibly expanded. Zeller's analysis of these points of divergence provides a roadmap for how to navigate climate cultures concerning geoengineering to address political issues slowing climate change adaptation and mitigation.

Next, sociologist Michael Haedicke's chapter theoretically expands and empirically applies the framework of climate cultures to coastal Southern Louisiana in the United States, a region environmentally ravaged by disasters, pollution, and degradation linked to climate change, industry, and conservative politics. Here, officials rely on "The Louisiana Master Plan for a Sustainable Coast", developed after Hurricane Katrina, to inform decisions and distribution of resources meant to boost the area's resiliency. Based on interviews with thirty-seven elite experts from business, politics, science, and civil society, Haedicke examines the opposing accounts of both supporters and critics of the plan, illustrating key disagreements over the best overall strategy (fortification versus relocation), the role and responsibility of the oil and gas industry, and the quality of the scientific research underlying the plan. In the respective climate cultures of supporters and critics, economic development agendas clash with social equity ones, and pro-business sentiment collides with calls for industry accountability. Overall, the chapter demonstrates that, and how, climate cultural struggles are linked to broader institutional and place-based inequalities. Haedicke concludes that a clear focus on power and conflict, favored by the political economy framework, will further strengthen existing approaches to climate cultures.

Lastly, sociologist Julia Teebken closes the section, and the entire book, by offering readers a fascinating case study of climate change activism, policy-making, and public debate under hostile conditions. Teebken's research consists of thirty-one qualitative interviews with pro-environment political actors, as well as participant observation of the first two Climate

Conferences in 2016 and 2019 in Georgia, a US state where climate skepticism has dominated political opinion and policy-making in past decades. The author examines how local policy practitioners have learned to adapt to the particular political environment, by developing an array of "coping mechanisms" for the initiation of climate change-related planning and/or policies – such as employment of non-climate frames, use of audience-specific language, or issue-linking to culturally prevalent topics. In doing so, Teebken highlights the interactive nature of political work in which actors must become highly aware of, and carefully adjust to, the political agendas of "the other". In sum, the author shows us that successful climate cultural discourse and action must be placed within the context of surrounding social and political "landscapes" in which they emerge and take place.

In closing, while the nine chapters of the volume are quite different in thematic focus, theoretical background, research location, analytic scope, and writing style, taken together, they particularly highlight two issues, in our view. First, the chapters demonstrate the current complexity and diversity of climate cultures on both sides of the Atlantic. Diverging, and often conflicting, climate cultures can be found in mass media, social media, professional, and policy spheres, as well as many other niches of the social world. Climate cultural beliefs and practices unite, and often divide, community members and social movement activists, just as they unite and divide policymakers, technical experts, and scientists. And second, as already mentioned above, the chapters in this volume illustrate the potential of the "climate cultures" concept and theoretical framework in charting discourse and action beyond the typical analytic levels of global hemispheres nations, or regions. It provides new puzzles and questions for future researchers who seek to examine the relational networks and spaces that define climate cultures with their global and local constellations in the 21st century.

Note

1 We sincerely thank Hadi Khoshneviss for alerting us to this funding opportunity.

References

Barnes, Jessica, and Michael R. Dove, eds. 2015. *Climate Cultures: Anthropological Perspectives on Climate Change*. New Haven, CT: Yale University Press.

Bembnista, Kamil, and Thorsten Heimann. 2020. "Zur diskursiven Konstruktion des Erinnerns: Resilienzkonstruktionen in öffentlichen Medien und bei Bewohnern in Hochwasserquartieren 20 Jahre nach der Oderflut von 1997" [On the Discursive Construction of Remembering: Resilience Constructions in Public Media and Among Residents in Flood Areas 20 Years after the Oder Flood in 1997]. In *Katastrophen zwischen sozialem Erinnern und Vergessen* [Catastrophes between Social Remembering and Forgetting], edited by Michael Heinlein, and Oliver Dimbath, 21–49. Wiesbaden (Germany): Springer VS.

Carvalho, Anabela. 2007. "Ideological Cultures and Media Discourses on Scientific Knowledge: Re-reading News on Climate Change". *Public Understanding of Science* 16 (2): 223–243.

Christmann, Gabriela, and Oliver Ibert. 2012. "Vulnerability and Resilience in a Socio-Spatial Perspective: A Social-Scientific Approach". *Raumforschung und Raumordnung* 70 (4): 259–272.

Christmann, Gabriela, Karsten Balgar, and Nicole Mahlkow. 2014. "Local Constructions of Vulnerability and Resilience in the Context of Climate Change. A Comparison of Lübeck and Rostock". *Social Sciences* 3 (1): 142–159.

Christmann, Gabriela, and Thorsten Heimann. 2017. "Understanding Divergent Constructions of Vulnerability and Resilience: Climate Change Discourses in the German Cities of Lübeck and Rostock". *International Journal of Mass Emergencies and Disasters* 35 (2): 120–143.

Crook, Tony, and Peter Rudiak-Gould. 2018. *Pacific Climate Cultures: Living Climate Change in Oceania*. Berlin (Germany): De Gruyter.

Douglas, Mary, and Aaron Wildavsky. 1982. *Risk and Culture: An Essay on the Selection of Technical and Environmental Dangers*. Berkeley and Los Angeles: University of California Press.

Dunlap, Riley E., and Aaron M. McCright. 2010. "Climate Change Denial: Sources, Actors and Strategies". In *Routledge Handbook of Climate Change and Society*, edited by Constance Lever-Tracy, 270–290. London (United Kingdom): Routledge.

Dunlap, Riley E., and Kent D. Van Liere. 1977. "Land Ethic or Golden Rule: Comment on 'Land ethic realized' by Thomas A. Heberlein, JSI, 28 (4), 1972". *Journal of Social Issues* 33 (3): 200–207.

Evans, David M., Alison L. Browne, and Ilse A. Gortemaker. 2020. "Environmental Leapfrogging and Everyday Climate Cultures: Sustainable Water Consumption in the Global South". *Climatic Change* 163 (1): 83–97.

Ford, Allison, and Kari M. Norgaard. 2020. "Whose Everyday Climate Cultures? Environmental Subjectivities and Invisibility in Climate Change Discourse". *Climatic Change* 163 (1): 43–62.

Goodman, Michael K., Julie Doyle, and Nathan Farrell. 2020. "Practising Everyday Climate Cultures: Understanding the Cultural Politics of Climate Change". *Climatic Change* 163 (1): 1–7.

Heberlein, Thomas A., and Bo Shelby. 1977. "Carrying Capacity, Values, and the Satisfaction Model: A Reply to Greist". *Journal of Leisure Research* 9 (2): 142–148.

Heimann, Thorsten. 2019. *Culture, Space, and Climate Change: Vulnerability and Resilience in European Coastal Areas*. New York: Routledge.

Heimann, Thorsten, Anna Barcz, Gabriela Christmann, Kamil Bembnista, Petra Buchta-Bartodziej, and Anja Michalak. 2021. "Vulnerability and Resilience Embedded in Discourses: Literature, Media, and Actors' Cultural Knowledge in German and Polish River Regions". *Space and Culture* 24 (3): 1–15.

Heimann, Thorsten, and Bishawjit Mallick. 2016. "Understanding Climate Adaptation Cultures in Global Context: Proposal for an Explanatory Framework". *Climate* 4: 59–71.

Hoffmann, Andrew. 2015. *How Culture Shapes the Climate Change Debate*. Stanford, CA: Stanford University Press.

Hulme, Mike. 2016. *Weathered: Cultures of Climate*. London (United Kingdom): Sage.

Kollmuss, Anja, and Julian Agyeman. 2002. "Mind the Gap: Why Do People Act Environmentally and What Are the Barriers to Pro-Environmental Behavior?" *Environmental Education Research* 8 (3): 239–260.

Liu, Wenling, Gert Spaargaren, Nico Heerink, Arthur P. J. Mol, and Can Wang. 2013. "Energy Consumption Practices of Rural Households in North China: Basic Characteristics and Potential for Low Carbon Development". *Energy Policy* 55: 128–138.

Nash, Nick, Lorraine Whitmarsh, Stuart Capstick, Valdiney Gouveia, Rafaella de Carvalho Rodrigues Araújo, Monika Dos Santos, Romeo Palakatsela, Yuebai Liu, Marie K. Harder, and Xiao Wang. 2020. "Local Climate Change Cultures: Climate-Relevant Discursive Practices in Three Emerging Economies". *Climatic Change* 163 (1): 63–82.

Shove, Elizabeth, Mika Pantzar, and Matt Watson. 2012. *The Dynamics of Social Practice: Everyday Life and How It Changes.* London (United Kingdom): Sage.

Smit, Mattijs. 2015. *Southeast Asian Energy Transitions: Between Modernity and Sustainability.* Farnham (United Kingdom): Ashgate.

Szerszynski, Bronislaw, and John Urry. 2010. "Changing Climates: Introduction". *Theory, Culture and Society* 27 (2–3): 1–18.

Wood, Donald C. 2015. *Climate Change, Culture, and Economics: Anthropological Investigations.* Bingley (United Kingdom): Emerald Group Publishing.

Part II
Contexts

1 Climate Cultures[1]

*Harald Welzer, Hans-Georg Soeffner
and Dana Giesecke*

Introduction

In early 2007, the Environment Board of the United Nations alarmed the global public by releasing three reports, stating that the climate system will collapse if CO_2 emissions continued as they were. However, the initial anxiety over what the scientists had announced did not endure because, among other reasons, it quickly became obvious that the consequences of global warming would be distributed extremely unevenly. While the countries of the Global South were destined to suffer droughts, floods, soil erosion, and other disasters in the future, there could well be some positive impacts in the rich countries of the Global North – on tourism, agriculture, and industry, provided that the latter engaged sustainable technologies.

The bad news by the *Intergovernmental Panel on Climate Change* (IPCC) initially caused considerable dissonance in people's minds; however, this concern was successfully reduced by applying familiar frames of perception and strategies of problem-solving, or rather non-solving. In the wake of the financial and economic crises of 2008 and 2009, the topic was permanently relegated to the second tier of attention. Today, in retrospect, one can conclude that while awareness of the problem has increased, the (un)willingness to change anything has remained constant. This finding, unsurprising for mediated societies, is primarily owed to the fact that the sheer impact of the problem of climate change on our future social and cultural life has not been assessed at all. Most people still seem to believe that the Western lifestyle, with its careless use and waste of external resources, can be globalized, and that in 20 years' time, it will still be possible to indulge in the same culture of consumption and leisure that is widely enjoyed today.

Climate change as uninterpreted cultural change

In many respects, climate change is an underestimated, and to date even largely unacknowledged, social danger – moreover, at present, it is unclear whether democratic societies are at all able to initiate the changes that are absolutely needed to avert the danger, or rather adapt to its

DOI: 10.4324/9781003307006-4

consequences. This doubt applies to all economic and social questions resulting from the dual stress of resource shortages on the one hand, and emission increases on the other, as well as the explosive questions of generational injustice, or of competition for resources, and the associated security issues.

It is entirely possible that, due to unmitigated climatic changes, the unexpected and uneven ascent of humanity has reached a negative dynamic, leaving forms of perception, interpretation, and problem solving that have developed over decades and centuries to lurch behind. The widespread inability to understand the proper scale of the global threat speaks to this, as does the common indifference to the violence that is factually and potentially caused by climate change. Besides, of course, from a global perspective, there are completely incompatible interests that will inhibit a determined, joint effort to slow the speed of global warming, even within a medium time frame. The catch-up process of industrialization in developing countries, the continuing hunger for energy in countries that have industrialized earlier, and the global spread of a societal model based on growth and resource depletion make it seem unrealistic that a global temperature increase of two degrees can be averted by the middle of this century. Moreover, this is a prospect based only on a linear view of growth, without considering auto-catalytic processes that may further accelerate the social impacts of climate change and escalate violence.

At the geophysical level, non-linear processes could occur that would radically exacerbate the climate problem – for example, when the thawing of permafrost releases enormous quantities of methane that will, in turn, impact the climate, or when forest loss and ocean acidification reach critical levels and generate as yet unforeseeable domino effects. The same applies to the social level – for example, when competition for natural resources triggers conflicts that, in turn, trigger waves of refugees, further intensifying border conflicts and piracy. All this can lead to unpredictable violence both within and between states. The logic of social processes is not linear; the same applies to the consequences of climate change. Nothing in the history of human violence suggests that longer periods of peace indicate permanently stable social orders; instead, the past proves that massive use of violence is *always* a possible course of action.

At present, there are deepening asymmetries in the global balance of power, just as there are wars caused by climate change that may lead to new cycles of unceasing violence. Because the most severe climate impacts will be shouldered by societies with the least coping capabilities, global migration will only increase throughout the 21st century, prompting those societies in which immigration is perceived as a threat to adopt radical solutions to the problem. It fundamentally appears to surpass our imagination that a phenomenon described as *natural*, such as global warming, encompasses social catastrophes, collapsing societies, civil wars, and genocides. However, not much imagination will be needed to grasp this fact, since violent environmental conflicts and heightened security measures can already be observed in the present.

Nonetheless, climate change has thus far been regarded as a phenomenon that falls into the purview of natural sciences, while members of the social and cultural disciplines and the humanities have taken, at best, a private, rather than a professional, interest in global warming. This happens despite the fact that meteorologists, oceanologists, paleoarcheologists, and glaciologists – in a rare agreement that, in some cases, goes back several decades – have confirmed not only that the average global temperature is indeed rising but also that human emissions activity, especially CO_2 emission, is largely responsible for this increase. Here, a particular complication is that the causes of the problem currently emerging date back at least half a century and could never have been anticipated based on the state of scientific research at the time. The whole issue becomes even more complex when considering that current intervention strategies aimed at the consequences of past actions – which could also not have been anticipated at the time – only carry a highly uncertain probability of success and, moreover, may only gain traction in a temporally distant future. In this case, the temporal link between actions and their consequences stretches across generations and, moreover, can only be comprehended through scientific mediation. It can hardly be grasped through experience. This makes it easy to simply ignore the problem, or to postpone serious solution attempts to a fictitious "later".

Just like mental dispositions, material and institutional infrastructures are inert, and the task of changing them is on a par in magnitude with mastering the first industrial revolution. At the same time, ambitions for growth and modernization in developing and newly industrialized countries complicate the transition to a post-carbon era: what is saved in emissions in one region of the world is lost again by economic growth in another. For this reason, despite the Kyoto Protocol, global emissions increase every year, and with each passing year, it becomes more difficult to reduce the total output. In the case of anthropogenic climate change, we are dealing with "first-time consequences" (in German *Konsequenzerstmaligkeit*) in the sense of Arnold Gehlen – meaning an event that overwhelms traditional reference frames for problem perception and solution attempts already because it is, in many respects, without precedent.

Nonetheless, and this is already discernible today, the consequences of global warming will drastically change conditions for human life and survival. Although this change occurs with highly variable regional intensity, it will undoubtedly have extremely massive economic, political, and mental consequences. Consider the implications of anthropogenic climate change for democratic nations: what does the disintegration of temporally intelligible cause-and-effect chains mean for the development of political consciousness, and for political decision-making, in general? Moreover, which impact does the lack of accountability inherent to this disparate process have on perceptions of emerging social consequences and potential solutions for climate change? In a few years' time, which solutions and political options will be considered possible that still seem quite unthinkable today?

Of course, an earth and climatic system, in whatever material form, will continue to exist in the future, regardless of whether earth warms or cools two, four, or eight degrees on average. Evolution as such proceeds without a sense of "value", it is completely dispassionate about change. Only the emergence of a species capable of designating different time zones and of separating present from past and future, brings about a distinction between what is unchangeably given, what can be accomplished through action in the present, and what is desired or feared in the future – a distinction between what is and what should be. Prehuman evolution does not know this distinction, it has and needs no ethics. Neither is a lion that kills a gazelle a murderer, nor are methane-emitting herds of buffalo or cattle committing a climate sin. In contrast, the suffering and death caused by humans can never be considered neutral in value.

A species that, on the one hand, knows that it, too, is part of "nature" but that, on the other hand, is characterized by "natural artificiality" (Helmuth Plessner) neither acts entirely instinctive nor in a purely reactive manner. The, in contrast to other living beings, extensive instinctual plasticity of man,[2] his "non-fixity", make him both a "specialist of non-specialization" and a "risk-taker" (Konrad Lorenz). He has no natural habitat but must build his own: the environment that suits man, "his" world, is artificial. It consists of what he attempts to create out of nature and also on his own: culture. For this culture, his world, he bears responsibility. Admittedly, on the one hand, no ethics of the world can disregard the fact that we, whether we want to or not, often (must) act irresponsibly and only recognize our irresponsibility in retrospect. On the other hand, we cannot take "personal" responsibility for everything and everyone. Neither can any pragmatics absolve itself from accepting the consequences of its actions, nor from having to justify them ethically.

As cultural and temporal beings, we are characterized far more by anticipatory adaptation than by reactive assimilation. While we are, and remain, part of an all-encompassing evolution that has produced, among other things, our future-oriented adaptive behavior, we also know that we must pay for this evolutionary gift. On the one hand, the experience of our natural artificiality forces upon us an understanding of finitude, weakness, and defectiveness, and with it a constant reminder of the need for improvement; on the other hand, as we try to take control, we realize that we ourselves are controlled. With the emergence of the human species (an "artificial" species, in a dual sense), evolution has – metaphorically speaking – given itself a reflexive component in man as a driven driver, and only in him. Man possesses both a future-oriented instrumental drive and a reflexive impulse to justify anticipatory measures of adaptation. And the more our capacity for control increases, the greater the pressure, on the one hand, to grow, to accept responsibility for reducing suffering, and on the other hand, to justify our options and choices. As a result of these processes of mutual reinforcement, two options are closed forever: standstill and going back in time. Therefore, for man, climate and culture are inextricably linked, as are climate change and cultural change.

Only within the framework of such an understanding of culture can techniques be developed by which measurements of past conditions allow predictions of future situations. This epistemological aspect alone renders climate change a legitimate object of study for the cultural disciplines. "Nature" is necessarily indifferent to whether humans exist or not, only the reverse is not true: climate change, in particular, highlights the extent to which human survival depends on the presence of favorable climatic conditions – regarding biological heat transfer, food options, energy supply, but also the many luxuriating aspects of human life that no longer have anything to do with sheer subsistence.

In 2005, within a few hours, Hurricane *Katrina* caused a complete collapse of the social order in a major city of the world's largest industrialized nation, and any winter storm in Central Europe can easily and instantly crumple entire transportation systems. The heat wave that hit Central Europe in 2003 claimed 30,000 lives. In 2009, due to warming temperatures, 305 cases of tick-born encephalitis were reported to the German *Robert Koch Institute* in regions and counties where such infections were unthinkable just a few years earlier. Spending for coastal and flood protection is just as affected by climate change as are fishery or viticulture. Because there is hardly an area of social reproduction left untouched by the consequences of global warming, Nicholas Stern, the former chief economist of the World Bank, calculated in 2007 that the economic cost of climate change amount to about one fifth of the global GPD (Gross Domestic Product). From today's point of view, we can barely anticipate what this means, for example, for the social security systems of industrialized welfare states, or for the design of social service institutions in developing and recently industrialized nations. The same applies to shifts in geopolitics and resource options, and for newly emerging hotbeds of violence.

Climate change as interpreted change: climate cultures

These few comments may already suffice to indicate that the phenomenon of anthropogenic climate change urgently calls upon the expertise of the humanities and cultural disciplines, starting with the question under which historic and cultural circumstances the interpretation of such a phenomenon is taking place to begin with. This scholarly expertise concerns historical knowledge of anticipated, witnessed, or experienced catastrophes, as well as their corresponding frames of interpretation. Further, it refers to the cultural practices and contexts of meaning that have led to the emergence of anthropogenic climate change, as well as extending to the broad matter of its social, political, psychological, and legal response. Last but not least, it challenges the human potential for interpretation and meaning making: the philosophical analysis of justice and responsibility, the philological or literary critique of language, as well as the sociological study of collective forms of knowledge and interpretation.

From this angle, it is obvious how glaring the neglect of the humanities and cultural disciplines has been, having left the topic of climate cultures largely unexamined – and they are pressed for an explanation of why this has happened. In our view, the biggest reason for the withdrawal of the social and cultural sciences, and the humanities, from reality-driven theorizing and, above all, from public and political discourse – something which, by the way, can be observed in many other respects as well – lies in the systemic collapse of the Eastern Bloc in 1989. Not only was this major shift in the global figuration of powers not anticipated by anyone, not even by the interpretive sciences actually in charge of these issues; but, for many colleagues, this event also invalidated previously accepted theories in their respective disciplinary traditions – whether they were of a Marxist or systems-theoretical shading. In more than a few sociology and political science seminars, from summer semester 1990 onwards, one no longer read Marx but Weber. And, as we know from looking back at the past two decades in these disciplines, it was not the only *turn* to occupy the humanities and cultural disciplines since.

"Discursive", "iconic", "visual", and "narrative", among others, are the names of other turns that followed one another without fail and that have succeeded, above all, in one regard, aside from bringing a certain theoretical sterility and a pervasive distance from the empirical world: they have pushed the subject matters of the humanities and cultural disciplines further and further away from the realm of social problems into the esoteric world of discourse. It is precisely this self-sufficiency of scholarly life in the world-free spaces of sheer and self-satisfied intellectualism that has led to the loss of critical potential in the social sciences, humanities, and cultural disciplines. Moreover, their ability to transcend the obvious has disappeared – as reflected, among other things, in the fact that they have lost the future and with it, necessarily, the fundamental care for their own and for a collective existence.

Through this obliviousness to the future, the social and cultural sciences and the humanities have contributed substantially to the depoliticization of the public sphere. When interpretive elites abandon their critical capability, democracy is deprived of a powerful corrective, and civil society of an analytic and therefore political force. The political emptying of the public sphere is not replenished, but rather exemplified, by a post-democratic simulation of political debate of the sort featuring in "Anne Will" or "Hart aber Fair".[3] The phenomenon of climate change has remained dramatically under-interpreted in terms of its social and cultural implications precisely because it was abandoned to the banter of empty phrases in talk shows and parliamentary debates.

The systematic negligence in the cultural sciences of a phenomenon that vastly contributes to determining the conditions for human life in the 21st century is fatal, not only because the social and cultural sciences and the humanities deprive themselves of a topic of profound heuristic value through their indolence regarding the unprecedented consequences of climate change. Above all, this is the case because not only the empirical study but also the communication of the phenomenon is left almost exclusively to colleagues from the natural sciences who are confined by their specific disciplinary perspectives. We can see this, for instance, in the entirely abstract discussion of the consequences of climate change. The information that the CO_2 content of the atmosphere will grow to 400 parts per million by the year 2020, or that sea levels will have risen by up to 89 centimeters at the turn of the next century, almost entirely obscure what this means for the "everyday life" of humanity, both globally and regionally.

We simply must describe the implications that these developments have for the living conditions and survival of humans. The natural sciences do not offer a scholarly account of this issue, and it does not fall within their responsibilities. But at the very moment, their findings on rising temperatures, melting ice sheets, or northward spreading malaria infections are discussed in public and political arenas, and questions about potential circumventions, adaptations, and solutions necessarily arise. Naturally, our colleagues in the natural science feel compelled to answer. Their answers, however, turn out to be decidedly naïve – such as appeals to behavioral changes – or technocratic, suggesting that geo-engineering, CO_2 storage, and electric cars will do the trick. Inevitably, such "advice" is insufficient with regard to the cultural disciplines. By definition, natural science and technological discourse about climate change is unable to ask questions about a historically grounded critique of technology; about economic and environmental history; about the origins of material, institutional, and mental infrastructures; about interests, agendas, and strategies; about social dynamics and unintended consequences of action; about path dependencies, cultural obligations, collective thinking, and so on. This is not the fault of natural and technical science scholars, but rather of those in the social and cultural sciences, and the humanities who have, thus far, largely refused to provide any expertise on these and many other aspects of the phenomenon of climate change.

Concluding thoughts

Just as climate change confronts us, again, with the idea of finitude in a culture in which availability and consumability of resources are perceived as infinite, the categories of past, present, and future are becoming surprisingly confused. It is possible that progress has ended quite a while ago, and that the future lies in returning to a time before developments that have failed or gone wrong – an operation that is completely alien to modernity.

This is one of the reasons why this book[4] begins with a future that has been lost and ends with a past that has become precarious. Precisely such findings illustrate how fundamentally important the contribution of the social and cultural studies and the humanities is to research on the impacts of climate change, and how inescapably massive the damage will be if we surrender its analysis, interpretation, and prognosis exclusively to our colleagues in the natural and technical sciences.

Notes

1 This text is a shortened, and very slightly adapted, translation of the introductory essay by Harald Welzer, Hans-Georg Soeffner, and Dana Giesecke published in 2010 in their edited book "KlimaKulturen: Soziale Wirklichkeiten im Klimawandel" (in English: Climate Cultures: Social Reality and Climate Change), pp. 7–19, Frankfurt am Main: Campus. The volume was a commemorative Festschrift for Claus Leggewie on the occasion of his 60th birthday.
2 The original German text speaks of "man" in the general sense of "human" and not only in relation to the male sex. To remain close to the original, we kept the male singular in the translation but, of course, we also intend to refer to both men and women.
3 In English, *Hard but Fair*. These are names of popular German TV talk shows around 2010.
4 The authors here refer to the book they edited in 2010.

2 Excavating Transatlantic Climates

An Archeology of Climate Discourse between Germany and the United States

Lawrence Culver and Doris McGonagill

Introduction

Climate has fascinated intellectuals and laypersons alike on both sides of the Atlantic. They not only recorded weather observations and utilized the evolving natural sciences to comprehend climate but also depicted and speculated about climate in correspondence, travelogues, literature, and art. From the Enlightenment to the present, this was a shared transatlantic climate discourse with ideas traveling west and east, though shared did not necessarily mean similar. Europeans began with the accumulated climate knowledge of European history, from ancient theories of climate to long memories of extreme weather events. Immigrant North Americans carried that knowledge but were new arrivals in an unfamiliar climate.

The climate of North America represented both promise and problem for European settlers and their descendants. Their evolving conceptualization of climate was rooted in prior European climatic cultural knowledge and expectations, the immediate need for personal, family, or community survival, and the longer-term desire for success, both in terms of profit and on the scale of the United States as a national project. What that project would become, however, was a core point of contention between divergent regions, economic interests, and political ideologies in a contentious republic where climate served as a proxy for many other interests and ambitions. Even after the Civil War, and the consolidation of federal authority over a continental nation, climate remained a concern. A key contradiction was that even as climate science advanced, and the scientific apparatus of the state grew exponentially, denial and fiction endured, and even prospered. Fact could exist alongside fabulation, a core contradiction that lingers in US climate culture, and helps explain why a vocal minority still refuses to accept anthropogenic climate change. Only much later and in intensive transatlantic dialogue, a mindset would develop that began to see limits to natural resources and viewed the natural world as a place of beauty imperiled by human action. Over time, new climate-cultural constellations would evolve that both transcended national borders and diversified existing discourses

DOI: 10.4324/9781003307006-5

and networks in both the United States and Europe (Heimann 2018; Heimann and Mallick 2016). Yet, these relatively recent formulations of climate knowledge grew from far older concepts and discourse.

This chapter attempts to develop an archeology of contemporary climate cultures by tracing the early evolution of transatlantic climate-change discourses from a humanities perspective. Engaging in a historical and literary cultural studies approach, this chapter brings together critical readings of historical, scientific, and literary texts and images, joining previously divided fields of study focusing *either* on the North American *or* the European context. The time frame of this chapter extends from the 17th to the late 20th century, with an emphasis on German-North American knowledge exchange during the 19th century. In seven segments, the chapter addresses early European settlement in North America and initial adaptation of traditional climate theories, the advent of science and its entanglement with politics during a period of new explorations and surveys undertaken by European powers and the new United States around 1800, the crisis of the "Year without a Summer" (1816), Manifest Destiny and the Antebellum period, the Civil War period and late 19th century, the contradictions of the 20th century, and important developments after the Second World War.

Early European settlers in North America and European climate theories

The first English settlements in North America proved disastrous. Roanoke was lost, its fate unknown. In the years after its founding in 1607, Jamestown suffered an 80% fatality rate, with likely episodes of cannibalism. Only later would tobacco and slavery provide an economic foundation for the colony. Company investor and settler expectations, inflated by Spanish imperial success, imagined the presence of large Native cities, like those in Mesoamerica. They further hoped that gold would be plentiful and that an easy passage to the Pacific lay nearby. They also made faulty predictions about the climate, forecasting the success of Mediterranean crops. These climatic expectations, while incorrect, were less fantastical. Instead, they were rooted in conceptualizations of climate dating to the classical world. For the ancient Greeks and Romans, climate was synonymous with geography. Climatic belts girdled the globe, from torrid regions at the equator to frigid ones in the Arctic. Expectations of a Mediterranean climate where orange trees would bloom were not irrational for a colony due west of Gibraltar, nor could island residents of a marine climate be expected to understand a continental climate, with much greater seasonal extremes in temperature. In response to a climate that did not match expectations, colonists could respond with either curiosity or wishful thinking. They would engage in both (White 2017; Zilberstein 2017; White 2015; Kupperman 1982; Glacken 1967).

European Enlightenment discourse embraced ancient climate theories, and weaponized them in new ways. Culturally determined climate zones had long been considered as corresponding with certain physical, intellectual, and cultural characteristics of the people living there. In the 18th century, climate theory – shot through with varying racist stereotypes – was used to justify contemporary colonial endeavors, from land appropriation to enslavement. However, in form of the Köppen classification, facets of the ancient theorems lived on well into the 20th century. One outspoken Enlightenment proponent was Johann Gottfried Herder (1744–1803). Herder's understanding of "clima" is significant for the formation of modern ecology, drawing on geography and biology, but also for the impact Herder's theories had in the North American context (Kirchhoff and Trepl 2009). The 18th-century polymath amalgamated natural and cultural history with aesthetic and teleological judgment, which connected particularly well with American exceptionalism. Hugely influential in the European cultural sphere, Herder's theories were reflected in the research programs of numerous academic disciplines, including classical geography and biology, that radiated from Europe and the United States.

Herder asserted a reciprocal relationship between climate – understood as physical conditions – and the "genius" or character of a people. According to Herder, people(s) were shaped by the climates they live in, but through cultivation they also shape these climates. Diverse climes and climates dictate unique forms of land use that require continuous adaptations, reinventions, and adjustments to form an "organic" functional unity. Over time, people(s) "acclimatize" to specific lands, an assumption that would prove especially influential in North America (Kirchhoff and Trepl 2009, 39 f.). Herder's theories were important for another reason. They are reflective of a paradigm shift with respect to wilderness and wastelands which – traditionally charged with negative overtones – were now redefined as gateways, created by God, to cultural self-perfection (Nash [1967] 2014; Kirchhoff and Trepl 2009). While the climates in such zones might be dangerous, Herder argued they were part of a latent pre-established harmony in divine creation that requires humans to cultivate to further the intended divine order, turn chaos into cosmos.

Of the political leaders and thinkers Americans would later revere as "founding fathers", Benjamin Franklin's interest in the natural world is best known, from his investigations of lightning and electricity to tornadoes. Franklin also was among the first to speculate about possible connections between the climate in Europe and North America. Following the eruption of the Islandic volcano *Lakagígar* in 1783, when tropospheric aerosols darkened the skies of the entire northern hemisphere, Franklin – at the time the American ambassador to France – held the dust clouds above the two continents responsible for the strangely discolored Parisian sky as well as a particularly frigid Philadelphia winter. German poet Johann Wolfgang von Goethe was to later describe the strange copper-tinged skies in his *Theory of Colors*, and the experience was to become a stepping stone for his climate studies. These individual instances of curiosity can be seen as preludes to the modern natural sciences and academic disciplines that would focus on studies of climate by the later 19th century.

Science, politics, and the era of European and US explorations around 1800

While few Americans were scientists, American colonists and settlers nevertheless attempted to determine climate by "reading" landscape. By looking at vegetation, soils, surface water, and other factors, they tried to determine if land would be useful for agriculture. They also believed that climate and human health were interconnected, with "miasmas" causing disease, and good air ensuring health (Valenčius 2002). The new nation exhibited similar preoccupations. After independence, Alexander Hamilton made a shrewd proposal that became the Northwest Ordinance of 1787. The states surrendered western land claims and, in return, the national government assumed their debts. From this point forward all new western lands absorbed in the United States would become federal land. The Northwest Ordinance delineated a system by which this land would be surveyed and sold. In terms of physical geography, this was one of the most lasting landmarks of the Enlightenment. A Cartesian grid spread across the US portion of North America, and in many places, the square mile survey sections are still visible from aircraft.

After Thomas Jefferson's gargantuan Louisiana Purchase in 1803, interest in the nation's new territorial holdings reached fever pitch. For much of the rest of the 19th century, exploring, surveying, and selling western lands would be a primary function of the national government of the United States. Indeed, before the Civil War, the sale of these lands was the primary source of revenue for the federal government. The government would fund a series of surveys of western lands. These surveys gathered valuable scientific and geographical data, but that was not their only purpose. Survey findings, bound in impressive volumes filled with maps, diagrams, analyses of soil and water, and attractive illustrations, were on the one hand compendiums of cartographic knowledge, and on the other hand glossy guides to new American real estate. In fact, the federal government spent more on these lavish publications than on the survey expeditions themselves (Culver 2012a; Goetzmann 1966).

Some of the surveys built on knowledge previously gathered by European explorers. Their observations and scientific notes offered important contributions to the evolving climate discourses both in North America and in Europe. In 1772–1776, Georg Forster (1754–1794), a German scientist, artist, and writer, participated in Captain James Cook's second expedition. The observations made on this *Voyage round the World*, published first in English, later in German, supplied a scientific, but also an aesthetic and emotional framework that made Forster an influential interlocutor for the emergence of a modern environmental imagination on both sides of the Atlantic (Wilke 2015).

When describing the polar regions, Forster presents scientific data (depth, temperature, saltiness of the oceans, the blueish or greenish color of the ice) and observations about atmospheric phenomena and animal population, but he also captures the conditions of whiteout, the experience of disorientation and loss of control, and aesthetic components of what has been dubbed the "artic spectacle" or "arctic sublime". Forster's descriptions inspired not only future scientists and explorers but also artists, among them Caspar David Friedrich, whose polar image *Das Eismeer* (The Sea of Ice, 1824, see Figure 2.1), in turn, shaped polar renditions of painters like Fredrich Church (*The Icebergs*, 1861) and Charles Raleigh (*Chilly Observations*, 1889) (Wilke 2015, 157–166).

French-German explorer, botanist, and novelist Adelbert von Chamisso (1781–1838) participated in a scientific voyage around the world aboard the Russian explorer ship *Rurik*. This expedition (1815–1818) discovered the North-West passage, and Chamisso was the first to map extensive segments of the Alaskan coastline. He described the lifestyle of the Inuit that allowed

Figure 2.1 Caspar David Friedrich, Das Eismeer, 1823–1824 (oil on canvas, 96.7 cm × 126.9 cm, Kunsthalle Hamburg, Hamburg, Germany).

them to survive in the Arctic climate and lend his name to the newly discovered *Chamisso Island*. Chamisso's diary of the expedition, *Reise um die Welt: Das Tagebuch 1815–1818* (1836), also compiled the first inventory of California's flora and fauna and described habits and lifestyle of the local people in the San Francisco Bay area, which his expedition visited in 1816 (Weinstein 1999).

Doubtlessly the most influential of the European explorers, however, was a "disciple" of Forster, German naturalist and explorer Alexander von Humboldt (Harvey 2020; Wulf 2015; Sachs 2006, 53). During a five-year expedition to Central and South America (1799–1804), Humboldt gathered a vast amount of data, including extensive geographical, geological, botanical, zoological, and climatological information, in addition to his wide-ranging anthropological, linguistic, and economical observations (Humboldt and Bonplant 1825). With state-of-the-art equipment that included barometers, thermometers, magnetometers, eudiometers, and dip needles, Humboldt traced the course of rivers, measured mountains, and observed environmental hazards related to vegetation, waterways, and air quality – climate factors we have come to think of as markers of the Anthropocene – both on a local and a global scale. In 1804, shortly after Jefferson had concluded the Louisiana Purchase, Humboldt visited the United States to pay his respect to the nation's third president and the American people "who understood the precious gift of liberty" (Wulf 2015, 96). Referring to himself jokingly as "half an American" on several occasions, Humboldt shared Jefferson's vision of an agrarian republic, and he was widely known and admired in the new nation (Wulf 2015, 98; Sachs 2006).

Humboldt also raised ecological issues, pointing out the environmental exploitation and devastation he had witnessed concomitant with colonialism (Schaumann 2017, 66). With a keen eye for humanitarian and societal injustices associated with the European settlers' practices in the New World, Humboldt decried the institution of slavery – inextricably linked to the United States' cotton production – as "the greatest evil" and a "disgrace", although he stopped short of addressing this politically sensitive subject with Jefferson directly (Wulf 2015, 197). However, during his six-week visit, Humboldt met many other leading figures of politics, arts, and culture, including James Madison, who was an avid admirer of Humboldt. In a famous speech delivered to the Agricultural Society in 1818, Madison echoed Humboldt's warnings about deforestation when describing the devastating ecological effects of Virginia's tobacco cultivation. Later, Madison's speech would be described as the beginning of American environmentalism (Harvey 2020).

In recent years, Humboldt has experienced a spectacular transdisciplinary revival. Hailed as a pioneering ecologist and scientific thinker, his environmental legacy and contribution to the transatlantic climate dialogue can hardly be overestimated (Schaumann 2017; Bühler 2016; Wilke 2015; Wulf 2015). Although Humboldt could not foresee the extent of anthropogenic climate change, his argument for the interconnectedness ("relations") of all organisms and systems has lost none of its potency (Sachs 2006, 49). Speculating about the possible effect of detrimental human interventions on the global climate, Humboldt, after his American travels, expanded his comparative climatology and began to examine worldwide climate patterns. In 1817, he introduced the graphic visualization of meteorological data in isotherms, lines on a map connecting geographical points with the same temperatures (Sachs 2006, 51). Vastly superior to traditional temperature tables, this representation made it easier to compare data and recognize global patterns. Isotherms allowed Humboldt to conceive of climate as a complex system of interdependencies between atmosphere, oceans, and continents, and made possible what Humboldt dubbed "comparative climatology" (Wulf 2015, 177 f.). In his *opus magnus, Kosmos* (or, in English, *Cosmos*, 1845–1862), Humboldt accurately described the natural greenhouse effect as a form of thermal radiation that impacts the very habitability of our planet. Half a century later, the calculations of Swedish scientist Svante Arrhenius, a distant relative of contemporary climate activist Greta Thunberg, showed how anthropogenic climate change impacts global temperatures (Holl 2019). Scientific research into climate change was propelled significantly by the traumatic experiences of the year 1816.

A watershed event: the "Year without a Summer"

While Americans were most interested in determining the climate in the territory of their growing republic, they were also interested in global climate events. One example, as mentioned above, was Franklin's interest in the year 1783; another was 1816, the so-called "Year without a Summer". The eruption of Mount *Tambora* on the Indonesian Island *Sumbawa* in 1815 – the most powerful volcanic eruption ever recorded – triggered floods, famine, and a wave of emigration across Europe. In a region still recovering from the Napoleonic wars, months of severe weather, including cold temperatures, persistent rain and hail, and lack of sunlight due to dust particles in the stratosphere, caused crops to fail and food prices to double. For 1816, the Central England Temperature (CET) record, collecting meteorological data since the mid-17th century, listed the coldest July on record. The profound impact of "Achtzehnhundert und erfroren" (meaning "1800 and frozen to death") was later commemorated by so-called *Hungertaler* – small coin-shaped metal containers similar to snuffboxes, containing pictures and texts detailing the suffering and listing the prices of basic foods – or by planting commemorative trees. Indirect consequences of the famine included the foundation of saving unions and, according to some accounts, the perishing of horses even led Karl Drais in 1817 to invent the *dandy horse*, the precursor of the bicycle (Gerste 2019, 197 f.).

The famine following the "Year without a Summer" also triggered a mass migration from Europe to North America, as well as outmigration from New

England to states and territories to the west, for the volcanic eruption meant that warm weather never arrived in North America, either. The climatic disturbance would also lead to a great deal of transatlantic correspondence as Americans and Europeans alike tried to understand the bewildering events, which also included strangely colored skies caused by volcanic ash, recorded in paintings by Caspar David Friedrich, John Crome, and J.M.W. Turner (Hubbard 2019). The cold, rainy summer of the subsequent year gave birth to one of the first and most famous works of science fiction, Mary Shelley's *Frankenstein; Or, The Modern Prometheus* (1818), while Lord Byron wrote the poem "Darkness" and the story "A Fragment", precursors and inspiration to Bram Stroker's *Dracula* (1897). The year 1816 is also considered one of the catalysts for Goethe's intense interest in meteorological phenomena (Wenzel and Zaharia 2012). These studies eventually culminated in the publication of a series of weather essays, atmospheric studies, and geological treatises that demonstrated complex interactions and correlations between different phenomena: air, water, clouds, mountains, and geographic conditions (Sullivan 2017, 2010; Tantillo 2002).

Manifest Destiny and the Antebellum period

For many Americans far removed from Napoleonic Wars and eager to spread westward, the "Year without a Summer" was a climatic hiccup in an inevitable march toward national greatness. By the early decades of the 19th century, the nation embraced the ideology of *Manifest Destiny*, the idea that white Protestant Americans were racially, politically, religiously, and economically superior to all other peoples, and that their advance across the continent as "the great nation of futurity" was foreordained (see Figure 2.2). The racism contained in this ideology was obvious. What has been less fully understood is the degree to which *Manifest Destiny* encompassed the natural world as well. Anglo-Americans did not merely expect the other inhabitants of North America to cooperate with their ambitions, they expected the same of the continent itself (Culver 2012a, 2012b; O'Sullivan 1839).

While the image of a female personification of America moving west across the continent, as settlers and technology follow and Native Americans and wild animals flee, is a florid illustration of *Manifest Destiny* as imagined in the Antebellum Period, it was not created until 1872, meaning after the Mexican and Civil Wars had been won, the Gold Rush had occurred, and the transcontinental railroad had been completed. It was, in fact, created as an image for a new tourist guidebook to a now firmly conquered US West. It depicted an ideology confidently claiming a future when, in fact, it looked backward at what could now be falsely remembered as an inevitable *fait accompli*.

Yet, it soon became apparent that the continent and its climates might not cooperate. Washington Irving, the popular author of the stories "Rip Van Winkle" and "The Legend of Sleepy Hollow", coined a term for much

Figure 2.2 John Gast, American Progress, 1872 (oil on canvas, 29.2 cm × 40 cm,
Autry Museum of the American West, Los Angeles, California, USA).

of the West: he called it a "Great American Desert", a wasteland destined to
always be inhabited by nomads (Irving 1836, 136). The unwelcome discovery
that much of the West was arid challenged the racial and religious notions
that undergirded *Manifest Destiny*. It also interfered with the goal of all the
exploring and surveying: applying labor to US nature and transforming it
into wealth, individual and national (Benedikter et al. 2015, 25). One response, though certainly not a common one, tried to embrace the desert as
central to America's magnificent wilderness, preparing the argument that
nature, as opposed to race or culture, was to define the nation's character
and its "rugged individualism". Henry David Thoreau even contended that
"in Wilderness is the preservation of the World" and wanted to set aside
"national preserves" for recreation (Wulf 2015, 295).

As already mentioned, the art of Caspar David Friedrich, one of Europe's
preeminent Romantic painters, was an inspiration for numerous American
artists, including Thomas Cole, Frederic Edwin Church, and Albert Bier-
stadt (Kirchhoff and Trepl 2009, 53; Novak 1995). Likewise, Humboldt's
concept of nature influenced politicians and transcendental poets, scientists, naturalists, and landscape painters alike. It also set the course for the
foundation of some of the country's most iconic institutions, such as the
Smithsonian, the Sierra Club, and the National Park Service. Explorers like
John C. Frémont (1813–1890), to whom the naming of the "Golden Gate"

strait is attributed, embraced the spirit of exploration and – steeped in Humboldt's writing – documented the fauna and flora of the "wild" American West (Sachs 2006, 50). Humboldt also became a mentor to American scientists like Louis Agassiz and Arnold Guyot, although their later nationalist and racist agendas were at odds with Humboldt's stance on racial equality and cosmopolitan concept of science (Harvey 2020, 17 f.).

Furthermore, Humboldt's concept of nature influenced American painters, including Frederic Edwin Church and the Hudson River School, and American writers from the transcendentalists to Edgar Allen Poe and Walt Whitman (Sachs 2006). Emerson declared Humboldt "one of those wonders of the world" (Harvey 2020, 15). Thoreau's classic memoir *Walden* (1854) owed a debt of gratitude to both Goethe's and Humboldt's works, which imparted on him visions of how scientific accuracy could be paired with poetic imagination. According to Wulf (2015, 257 f.), Humboldt's *Cosmos* was to Thoreau a healing "elixir", and "standing on the Concord cliffs", Thoreau was in spirit "with Humboldt" (Sachs 2006, 27). Likewise, Sierra Club founder John Muir was so riveted by Humboldt's writings that when the naturalist walked across the United States, channeling Humboldt's theories of unity and interrelatedness, he desired "to *be* a Humboldt" (Wulf 2015, 315, emphasis added).

The Civil War period and beyond

The immediate practical responses to reconcile the country's challenging geography and climate with the American assertion of a *Manifest Destiny* were different and more contentious. Decades before the Civil War, a regional split was already apparent in the United States, and it centered on interpretations of climate. Many Northerners embraced the idea of the Great American Desert, a wasteland that would halt the spread of slavery. Meanwhile, in the South, plantation owners were relentlessly intent on expanding their slave and cotton empire to the Pacific. Their climatic fixation had some origin in the fickle nature of their cash crop – cotton. Cotton is uniquely climate sensitive, requiring large and regular amounts of rain, and an average temperature no lower than 25°C/77°F, day and night. As a result, cotton could only be grown in the deep South, from the Carolinas through Georgia and Alabama, in the Mississippi Delta, and later in east Texas.

One of the Americans profoundly touched by Humboldt's insights into the ecological function of the forest was agriculturist and meteorologist Daniel Lee, editor for the *Southern Cultivator*, a magazine dedicated to agriculture and cotton cultivation in the Antebellum South. With a nod to Humboldt, Lee lamented in 1848 the excessive deforestation in parts of the country and drew attention to the "changing climate". Lee's reception highlights a problematic dimension of the entanglement between early climate science and economic interests: a supporter of slavery to achieve agricultural success,

Lee also thought that the South's reliance on slave-grown cotton at the expense of other crops was a problem.

Southern slave owners flatly refused to accept accounts of an arid, inhospitable West. Instead of the Great American Desert, they imagined a vast plantation garden. This vision would lead them into Mexican Texas, where there was indeed a good climate for cotton cultivation. The government of newly independent Mexico offered a vast land grant in eastern Texas to Stephen Austin, who led an influx of Anglo-Americans into the area, ignoring Mexican law which banned slavery. That, in turn, lead to an Anglo slave-owner revolt in Texas, and eventually its annexation into the United States. This prompted the US–Mexico War, and eventually the loss of more than half of Mexico's territory. In their moment of triumph, Southern slave owners would tolerate no bad news. A new government survey was created to delineate the new boundary between the United States and a diminished Mexico. When the leader of the boundary survey party, John Russell Bartlett, a New Yorker, reported that much of this land was barren, worthless desert, and that the new Southwest has not been worth a war, Southern members of the survey party rebelled. They complained to their members of Congress, and Bartlett was removed from command. He was replaced with William Emery, a Southerner who did not traffic in fictions but did engage in aggressive optimism, claiming in his report that while the Southwest was indeed more arid than the Southeast, through slavery, it could definitely prosper (Greenberg 2009).

By the middle of the 19th century, however, other mindsets were emerging. The same nation that produced climatic myths to spread slavery also witnessed the early stirrings of an environmentalist ethos. George Perkins Marsh, one of America's first conservationist and precursor to modern-day environmentalism, recognized the irreversible human impact on the climate and advocated sustainability concepts. In *Man and Nature* (1864), an early discussion of contemporary ecological problems that addresses water pollution, the creation of irrigation canals, deforestation, and desertification that went hand in hand with westward expansion, Marsh referenced Humboldt's *Ansichten von der Natur* (Views of Nature, 1808), an account of Humboldt's (South)-American observations, no less than fifteen times. (The title Marsh had originally considered for *Man and Nature* was *Man, the Disturber of Nature's Harmonies*. The revised edition of 1874 bore the title *The Earth as Modified by Human Action: Man and Nature*.) Marsh extolled Humboldt's scientific and philosophical credentials as well as his power of premonition when cautioning against excessive clearing of woodlands which might result in the diminution of water, rendering the land uninhabitable. Marsh's book imparted Humboldt's ideas to many leading figures of the preservation and conservation movement, including John Muir and Gifford Pinchot, and influenced both the Timber Culture Act of 1873 (aimed at environmentally reshaping the West by offering the Euro-American settlers deeds to public lands in return for growing trees) and the Forest Reserves Act of 1871 (Wulf 2015, 295 ff.).

Marsh's proto-conservation stance was an important influence on John Wesley Powell, a Civil War veteran, explorer of the Grand Canyon and Colorado River, and first head of the United States Geological Survey, who warned that eastern-style homesteading would not succeed in this new, harsher West. Instead, he urged the federal government to adopt the successful irrigation agriculture that made possible the Latter-day Saint (Mormon) settlement near the Great Salt Lake in 1847, and the irrigation and dry farming techniques practiced by Pueblo Indians and Hispano settlers in northern New Mexico and the valley of the Rio Grande. Furthermore, he argued for the vast expansion of homesteading tracts, and counseled settlers to pursue raising livestock for grazing rather than farming crops. He also believed that the regional irrigation culture in Utah would have to be scaled up to a vast federal water management system, building dams, reservoirs, canals, and irrigation projects. In essence, Powell was arguing for a hydrological reengineering of the West, one that could triumph over aridity (Worster 2002; Reisner 1986).

Congress and many homesteaders, however, ignored Powell. His plans would take time and cost money. Congress – not for the last time – when confronted with science it did not like, found science it did like. Railroad companies, which had been paid for railroad construction with government land, likewise had no interest in long-term expensive investments in land they wanted to sell rapidly. Instead, national economic ambitions led politicians, settlers, and railroad executives to seize upon a disastrous climatic myth: the claim that rain would "follow the plow". In this climate fantasy, plowing the earth would release moisture into the air. This would then fall as rain, which plowed earth more easily absorbed. This would achieve a permanent transformation of the climate. Of course, it was utter climatic fantasy, yet appealing nevertheless. The popularity of "rain follows the plow" also coincided with two decades of greater than average precipitation on the western plains. It initially seemed that the climate myth was proving true.

Fatefully, it was not. Drought and blizzards in the 1880s would wreak havoc on vast speculative cattle enterprises on the western plains, where millions of cattle died. Yet, the settlement movement continued. Huge numbers of European immigrants, who had no experience with the climate of North America, were lured by pamphlets making wild claims like one, printed in German, asserting that the valley of the Platte River in Nebraska was more fertile than that of the Rhine. As a consequence, unprepared settlers poured in (Wilber 1881).

In one of the deepest contradictions of climatic thinking in the United States, the disastrous adoption of climate fantasy occurred at the same time as the natural sciences began to apprehend the realities of climate, and the federal government of the United States rapidly developed the agencies and apparatus of state-sponsored science. Much of this US science spoke "with a German accent". A number of American pioneers of forestry education, management, and administration were German-born or German-trained. Bernard Fernow (1851–1923) is often considered the "father of professional forestry in the United States" (West 1999). Influenced by Swedish Enlightenment botanist Carl Linnaeus (1707–1778) and German silviculturist Heinrich Cotta's *Anweisung zum Waldbau* (Instruction in Silviculture, 1828), Fernow garnered support for a Prussian-style national forest service in the United States (Twight 1990). Another influential German-born forester was Carl A. Schenck (1868–1955). Formally trained at Giessen University, renowned for Justus Liebig's pioneering research on nitrogen-based fertilizers, Schenck had a distinguished foresting career working on the Vanderbilt estate in North Carolina, where he founded Biltmore Forest School. Finally, Gifford Pinchot (1865–1946), the first chief of the newly created US Forest Service (1905), corresponded with leading European forestry scholars, including Dietrich Brandis and Wilhelm Philipp Daniel Schlich. In sum, US climate discourse, early climate action, and the emergence of new disciplines, such as forestry, in the 19th century were heavily influenced by developments in the German and European scientific and academic arena, and transatlantic exchange and discourse concerning climate was widespread and influential.

Contradictions of the 20th century

By the 1900s, the US federal government had belatedly begun to adopt some of the ideas of John Wesley Powell. The Newlands Reclamation Act of 1902 funded the construction of dams and irrigation projects. The first was a dam on the Salt River in southern Arizona, which facilitated the growth of metropolitan Phoenix. Later, during the 1930s New Deal, the Army Corps of Engineers and Bureau of Reclamation would oversee vast dam building projects, like Boulder (later Hoover) Dam on the Colorado River, and all the dams on the Columbia River, radically transforming the hydraulic landscape of western North America. It seemed that big science and big engineering sponsored by a powerful national government had finally triumphed over the harsh climate of the American West and erased the Great American Desert once and for all. Now farms and cities could prosper.

Yet, this era also marked the arrival of a major new environmental voice. Aldo Leopold, the father of American wildlife ecology, developed his "land ethic" through a cross-cultural encounter with German models of land use and conservation (Leopold 1949). In a sometimes troublingly guileless exchange with preservation communities in what was by then Nazi Germany, Leopold cited what he perceived as the Germans' caring, nurturing relationship between people and nature. This relationship, in Leopold's perspective, transcended economic self-interest and was conscious of aesthetic and ethical obligations to the non-human world. Ignoring "völkisch" (nationalist) overtones of the unwholesome alliance between *The Green and the Brown* (Uekoetter 2006), Leopold identified a form of emotional community that was to ideologically underpin also much of the German environmental movement of the second half of the 20th century.

While the US victory in World War II, cementing the United States as a superpower for the remainder of the 20th century and unleashing the atomic age, suggested to Americans that, in some sense, *Manifest Destiny* had indeed come true, the dominant European and indeed German climate narrative, politically and environmentally, developed very differently. America had become the "great nation of futurity" on a scale that Antebellum Americans could scarcely have imagined. The irony of this is that in the post-war era, a population that had spent generations thinking, arguing, and worrying about climate stopped thinking about it almost entirely, at least as a matter of survival. With colossal dams and the advent of air conditioning, climate, which had once been a subject of dire concern for Americans, became simply a benign commodity or amenity, turning the hot summers of the Sunbelt from Florida to California into hospitable environments.

Yet, the Sunbelt migration, far larger than the homesteading migration had ever been, delivered huge numbers of American homeowners to places of dire climatic peril. While in the 19th century, the federal government of the United States tried to create successful farmers, it now wanted to create middle class suburban homeowners. With federal FHA and VA loans, they bought new homes in hurricane prone Florida and Texas, the drought prone Southwest, and fire prone California. This was, in its own way, a product of a newer kind of pernicious climate fantasy, one that imagined technology had rendered climate impotent. Not for nothing do Americans often call air conditioning "climate control" (Davis 1998).

By the 1970s, the American oil industry would actively combat and suppress research in climate science and the growing awareness of climate change among scientists. From Jamestown to Exxon, climate reality often succumbed to climate fantasy if there was profit at stake. This hard fact underlay all climate culture in the United States, even as it developed a robust and sometimes world-leading government-supported climate science connected to the larger scientific world. This core contradiction helps explain why, even in the 21st century, climate science in the United States struggles against wishful thinking, willful climate fantasy, and a refusal to believe "fake news". These contradictory beliefs contribute to the rise of diverse contemporary climate cultures in the United States (Heimann 2018).

The Post-World War II period

While in the 19th century, US climate science sometimes conflicted with ambitions for national territorial expansion and settlement, in Post-World War II Germany, science and scientists commanded a high level of respect since they were largely seen as representing German achievements supposedly unsullied by the crimes of the Nazi era. Many policymakers and scientists felt that when it came to contentious environmental topics like air pollution, water quality, and energy production, science popularization might be the best way to win over public support. In both East and West Germany, appealing to the "sensible public" was commonly part of the official strategy, whether in order to maintain the growth narratives of the German coal industry ("Saubere Kohle" or "clean coal"), the German car industry's "Freie Fahrt für freie Bürger" ("free passage for free citizens", one of the *Autobahn* nation's most popular slogans), or the promotion of "peaceful" uses of atomic power. Official GDR publicity enthusiastically embraced nuclear power. In the Federal Republic, however, official attempts of strategic science popularization had already begun to backfire by the late 1950s. After two nuclear accidents in 1958 at plants in *Kyshtym* in the Soviet Union and at the *Vinča* Nuclear Institute in Serbia (then Yugoslavia), popular West German media (including *Stern* magazine) began to depict nuclear power as a risky technology and even minimal exposure to radiation as harmful. This would be heightened greatly by the *Chernobyl* disaster (1986), after which the West German Green Party gained increasing popularity. And while the GDR authorities tried to smother public discourse, debates in the Federal Republic in the west trickled across the border, nourishing a tentative non-Communist ecological movement supported by parts of the East German Protestant Church in the east. Today, institutions like the *Deutsches Klima-Konsortium* (DKK, or German Climate Consortium) continue the post-war tradition of science-informed public relations work. Representing more than twenty renowned German climate and climate impact research organizations the consortium offers everything from free flyers and brochures to online courses on climate change and access to climate study data bases, as well as stipends and awards in climate-related professions and fields of study.

In the 1990s, key concepts and methodologies of ecocriticism were honed at anglophone universities, and, with some delay, this academic environmental discourse made its way from the United States and the anglophone world to Germany and Europe. But deteriorating air, soil, and water quality, and the necessity of renewable energy sources had been the subject of intense public discussion in Europe since the 1980s when *Saurer Regen* or Acid Rain, *Waldsterben* (forest death as a result of atmospheric pollution), and the nuclear disaster at *Chernobyl* (1986) threw into sharp relief a number of real and imminent threats to the environment. The smiling sun logo and the still popular slogan *"Atomkraft? Nein Danke!"* ("Nuclear Power? No thanks!"), originally designed by Danish student Anne Lund, came to emblematize the German, Scandinavian, and European anti-nuclear movement since the 1970s. "Green" initiatives like recycling, the use of solar panels, and air monitoring networks that provide up-to-date air quality data have been particularly successful in European countries that, since the formation of the European green parties in the 1970s, have had a strong representation of ecological and climate-related concern on the parliamentary level. These tend to be also the countries with the highest degree of awareness regarding the phenomenon of global warming.

A historical overview demonstrates that climate discussions and actions gain momentum with instances of objective environmental impacts or disasters (*Kyshtym*, *Vinča*, *Chernobyl*, and more recently *Fukushima*). Likewise, the accelerating frequency of catastrophic wildfires, floods, and hurricanes in the 2010s and early 2020s convinced a majority of US citizens of the reality of climate change. But increased societal perception and shifting discussion and information formats are also factors that accelerate the climate conversation. An example of such a shift is the contemporary youth movement spearheaded by Swedish climate activist Greta Thunberg. This movement

Figure 2.3 Caspar David Friedrich, Wanderer above the Sea of Fog, 1818 (oil on canvas, 94.8 cm × 74.8 cm, Kunsthalle Hamburg, Hamburg, Germany).

reflects and propels much of today's popular climate discourse, and in 2019, Thunberg was elected *Time* magazine's youngest Person of the Year ever, for her fight against climate change. Intriguingly, Thunberg's *Time* magazine cover included a not-so-subtle allusion to Caspar David Friedrich's *"Wanderer above the Sea of Fog"* (Figures 2.3 and 2.4), reminding us once

Figure 2.4 Cover of Time Magazine, 23 December 2019, showing Climate Activist Greta Thunberg, photographed on the shore of Lisbon, Portugal, 4 December 2019.

more how much of the contemporary explicit climate discourse is rooted in earlier cultural manifestations, part of a continuum, rather than something wholly new.

Conclusion

Today, the ideas of 1800 have come full circle, demonstrating once more how leading voices in the climate discourse – thinkers, artists, and scientists – in turn inspire each other, and how transatlantic climate conversations have been – and still are – evolving in dialogue and mutual stimulation. This, in turn, highlights the value of cultural and historical perspectives, for the diverse antecedents of contemporary climate culture shaped both cultural perceptions and science, and shape climate discourse still. In the past and present, scientific data and climatic events were perceived and comprehended through a lens of culture, and culture must be a component of successful climate change policy today. For while climate change is an environmental process, fundamentally it is a human problem. Understanding the history and significance of cultural and scientific climate discourse across the Atlantic prepares us to better understand contemporary discourses, developments, and divisions. Each of the chronologically or geographically focused studies in the following chapters trace specific examples that derive in part from an older, broader discourse, reminding us that 21st-century concerns about climate and climate change are connected to a deep and surprisingly diverse history of climate knowledge. Here, we have undertaken an archeological "excavation" of transatlantic climatic thought and dialogue and attempted to survey the cultural and rhetorical strata underlying contemporary scientific discourse.

References

Benedikter, Roland, Eugene Cordero, and Anne M. Todd. 2015. "The 'American Way of Life' and US Views on Climate Change and the Environment". In *Cultural Dynamics of Climate Change and the Environment in North America*, edited by Bernd Sommer, 21–54. Leiden: Brill.

Bühler, Benjamin. 2016. *Ecocriticism: Grundlagen – Theorien – Interpretationen*. Stuttgart (Germany): J. B. Metzler.

Chamisso, Adelbert von. [1836] 2015. *Reise um die Welt: Das Tagebuch 1815–1818*, edited by Michael Holzinger. Berlin (Germany): Holzinger.

Culver, Lawrence. 2012a. "The Desert and the Garden: Climate as Attractor and Obstacle in the Settlement History of the Western United States". *Global Environment* 5 (9): 130–159.

Culver, Lawrence. 2012b. "Manifest Destiny and Manifest Disaster: Climate Perceptions and Realities in United States Territorial Expansion". In *American Environments: Climate Cultures Catastrophe*, edited by Christof Mauch and Sylvia Mayer, 7–30. Heidelberg (Germany): Universitätsverlag Winter.

Davis, Mike. 1998. *Ecology of Fear: Los Angeles and the Imagination of Disaster.* New York: Metropolitan Books.

Gerste, Ronald D. 2019. *Wie das Wetter Geschichte machte: Katastrophen und Klimawandel von der Antike bis heute* [How the Weather Made History: Disasters and Climate Change from Antiquity to Today]. Stuttgart (Germany): Klett-Cotta.

Glacken, Clarence C. 1967. *Traces on the Rhodian Shore: Nature and Culture in Western Thought from Ancient Times to the End of the Eighteenth Century.* Berkeley: University of California Press.

Goetzmann, William H. 1966. *Exploration and Empire: The Explorer and the Scientist in the Winning of the American West.* New York: Vintage Books.

Greenberg, Amy S. 2009. "Domesticating the Border: Manifest Destiny and the 'Comforts of Life' in the US-Mexico Boundary Commission and Gadsden Purchase, 1848–1854". In *Land of Necessity: Consumer Culture in the United States–Mexico Borderlands*, edited by Alexis McCrossen, 83–112. Durham, NC: Duke University Press.

Harvey, Eleanor J. 2020. *Alexander von Humboldt and the United States: Art, Nature, and Culture.* Princeton, NJ: Princeton University Press.

Heimann, Thorsten. 2018. *Culture, Space and Climate Change: Vulnerability and Resilience in European Costal Areas.* New York: Routledge.

Heimann, Thorsten, and Bishawjit Mallick. 2016. "Understanding Climate Adaptation Cultures in Global Context: Proposal for an Explanatory Framework". *Climate* 4 (59): 1–12.

Holl, Frank. 2019. "War Alexander von Humboldt der erste Ökologe?". *Der Tagesspiegel.* 4 June. https://www.tagesspiegel.de/kultur/geschichte-der-klimaforschung-war-alexander-von-humboldt-der-erste-oekologe/24415586.html

Hubbard, Zachary. 2019. "Paintings in the Year without a Summer". *Philologia* 11 (1): 17–33.

Humboldt, Alexander von. 1845–1862. *Kosmos: Entwurf einer physischen Weltbeschreibung.* Stuttgart and Tübingen (Germany): J.G. Cotta'scher Verlag. 5 Volumes.

Humboldt, Alexander von, and Aimé Bonplant. 1825. *Personal Narrative of Travels to the Equinoctial Regions of the New Continent during the Years 1799–1804*, trans. Helen Maria Williams, Vol. 4. London: Longman.

Irving, Washington. 1836. *Astoria, or Enterprise beyond the Rocky Mountains.* Philadelphia, PA: Carey, Lea, and Blanchard.

Kirchhoff, Thomas, and Ludwig Trepl, eds. 2009. *Vieldeutige Natur. Landschaft, Wildnis und Ökosystem als kulturgeschichtliche Phänomene* [Ambiguous Nature. Landscape, Wilderness and Ecosystem as Cultural-historical Phenomena]. Bielefeld (Germany): Transcript.

Kupperman, Karen O. 1982. "The Puzzle of the American Climate in the Early Colonial Period". *The American Historical Review* 87 (5): 1262–1289.

Leopold, Aldo. 1949. *A Sand County Almanac: And Sketches Here and There.* Oxford (United Kingdom): Oxford University Press.

Marsh, George P. 1864. *Man and Nature; or, Physical Geography as Modified by Human Action.* New York: Charles Scribner.

Nash, Roderick F. [1967] 2014. *Wilderness and the American Mind.* 5th ed. New Haven, CT: Yale University Press.

Novak, Barbara. 1995. *Nature and Culture: American Landscape and Painting, 1825–1875.* New York: Oxford University Press.

O'Sullivan, John L. 1839. "The Great Nation of Futurity". *The United States Democratic Review* 6 (23): 426–430.

Powell, John W. 1879. *Report on the Lands of the Arid Region of the United States, with a More Detailed Account of the Lands of Utah.* 2nd ed. Washington, DC: Government Printing Office.

Reisner, Marc. 1986. *Cadillac Desert: The American West and Its Disappearing Water.* New York: Penguin Books.

Sachs, Aaron. 2006. *The Humboldt Current: Nineteenth-Century Exploration and the Roots of American Environmentalism.* New York: Viking.

Schaumann, Caroline. 2017. "Calamities for Future Generations: Alexander von Humboldt as Ecologist". In *Ecological Thought in German Literature and Culture*, edited by Gabriele Dürbeck, Urte Stobbe, Hubert Zapf, and Evi Zemanek, 63–76. Lanham, MY: Lexington Books.

Sullivan, Heather I. 2010. "Ecocriticism, the Elements, and the Ascent/Descent into Weather in Goethe's Faust". *Goethe Yearbook* 17: 55–72.

Sullivan, Heather I. 2017. "Goethe's Concept of Nature: Proto-Ecological Model". In *Ecological Thought in German Literature and Culture*, edited by Gabriele Dürbeck, Urte Stobbe, Hubert Zapf, and Evi Zemanek, 17–29. Lanham, MY: Lexington Books.

Tantillo, Astrida O. 2002. *The Will to Create: Goethe's Philosophy of Nature.* Pittsburgh, PA: University of Pittsburgh Press.

Thoreau, Henry D. 1854. *Walden; or, Life in the Woods.* Boston, MA: Ticknor and Fields.

Twight, Ben W. 1990. "Bernhard Fernow and Prussian Forestry in America". *Journal of Forestry* 88 (2): 21–25.

Uekoetter, Frank. 2006. *The Green and the Brown: A History of Conservation in Nazi Germany.* Cambridge: Cambridge University Press.

Valenčius, Conevery B. 2002. *The Health of the Country: How American Settlers Understood Themselves and Their Land.* New York: Basic Books.

Weinstein, Valerie. 1999. "Reise um die Welt: The Complexities and Complicities of Adelbert von Chamisso's Anti-Conquest Narratives". *The German Quarterly* 72 (4): 377–395.

Wenzel, Manfred, and Mihaela Zaharia. 2012. "Schriften zur Meteorologie" [Writings on meteorology]. In *Goethe Handbuch, Supplemente 2* [Goethe Handbook, Supplements 2], edited by M. Wenzel, 206–224. Stuttgart: J. B. Metzler.

West, Terry. 1999. "Fernow, Bernhard Eduard". *American National Biography* (online ed.). New York: Oxford University Press. https://doi.org/10.1093/anb/9780198606697.article.1000540

White, Sam. 2015. "Unpuzzling American Climate: New World Experience and the Foundations of a New Science". *Isis* 106 (3): 544–566.

White, Sam. 2017. *A Cold Welcome: The Little Ice Age and Europe's Encounter with North America.* Cambridge, MA: Harvard University Press.

Wilber, Charles D. 1881. *The Great Valleys and Prairies of Nebraska and the Northwest.* Omaha, NE: Daily Republican Print.

Wilke, Sabine. 2015. *German Culture and the Modern Environmental Imagination: Narrating and Depicting Nature.* Leiden/Boston, MA: Brill Rodopi.

Wood, Gillen D'Arcy. 2014. *Tambora, the Eruption that Changed the World.* Princeton, NJ: Princeton University Press.

Worster, Donald. 2002. *A River Running West: The Life of John Wesley Powell*. New York: Oxford University Press.

Wulf, Andrea. 2015. *The Invention of Nature: Alexander von Humboldt's New World*. New York: Alfred A. Knopf.

Zilberstein, Anya. 2017. *A Temperate Empire: Making Climate Change in Early America*. New York: Oxford University Press.

Part III
Europe

3 Capturing Climate-Cultural Diversity

A Comparison of Climate Change Debates in Germany

Sarah Kessler and Henrike Rau

Introduction

According to Germany's leading weekly newsmagazine *Der Spiegel*, a cleverly timed video by the German YouTuber *Rezo* that attacked the ruling Christian Democratic Union (CDU) party for its inaction concerning climate change seems to have contributed to making the Green Party the winners of the European election in Germany (Backes et al. 2019).[1] Rather unusually for videos aimed at Rezo's young target group, his clip titled "The Destruction of the CDU" is 55 minutes long and entails 13 pages of references. It was the most watched YouTube video of 2019 in Germany. While its impact on the outcome of the elections remains unclear, it evidently influenced subsequent climate debates in Germany. Moreover, its eminence reveals the strong and growing influence that prominent figures, here one belonging to the new influencer-generation, can exert on people's opinions regarding who should be held responsible for climate protection. The range of reactions incurred within its resonance also demonstrates the diversity of views that characterize climate debates in Germany. These heated debates illustrate the challenges faced by those interested in advancing climate protection, including the need to develop new concepts of knowledge that include both cognitive and emotional elements as "emotions are the conduit between mind (culture) and body (nature), and they define our first filter of interpretation of the external world" (Davidson 2018, 379).

Accordingly, this study captures the variations in people's perceptions of climate change, linking them to divergent and potentially conflicting "climate cultures" (Hulme 2017; Heimann and Mallick 2016): *group-specific variations in how climate change is viewed and dealt with in different social contexts.* Based on rich empirical data, a broad distinction can be made between *elite* climate cultures and climate cultures *"from below"*. The study thus moves beyond much existing research on public attitudes and understanding of climate change that focuses on the expressed views of individuals. Moreover, it extends existing work on climate cultures by defining them as *group-specific reference points for recognizing culturally relevant information regarding climate change that shape and reflect the climate-relevant*

DOI: 10.4324/9781003307006-7

everyday practices of their members, including people's lived experiences of responsibility and efficacy regarding climate (in)action. Climate cultures, therefore, constitute *shared repertoires of cognitive, emotional, and behavioral responses to the threat of climate change that characterize particular segments of society and that become visible in public climate debates,* including those examined in this chapter.

To identify different and potentially divergent climate cultures, this study examines statements made by different actors, including politicians, scientific experts, prominent public figures, and members of the general public within a particular cluster of media coverage. The sample consists of interconnected sources, namely three primetime talk shows broadcasted between March and June 2019, responses to these talk shows on social media, in print and online news, and a selection of YouTube videos posted by prominent influencers.

Mass media coverage is the main source of people's knowledge on many climate matters (Schäfer and Bonfadelli 2017). However, the social context within which the users consume and interpret climate-related media content remains seriously under-researched (Schäfer and Bonfadelli 2017). Exceptions include some previous social-scientific work on the discrepancies between how the public is imagined by political and scientific elites, and the characteristics of the actual public, that is, a socially embedded citizenry that holds different views and engages in a wide range of different practices (Fox and Rau 2017; Walker et al. 2010; Maranta et al. 2003). Here, contrasts between rather homogeneous and highly visible elite climate cultures and heterogeneous and less visible climate cultures "from below" stood out and therefore motivated this study.

Media attention to climate change remains strongly linked to meteorological and political events (e.g. COP meetings; Neverla et al. 2018, 23; Schäfer and Bonfadelli 2017, 11; Schmidt et al. 2013; Schäfer 2012, 8 f.). Resulting variations in attention, in turn, can significantly influence electoral outcomes, as has said to have been the case with the 2019 European election that provides the timeframe for this study's data collection. We argue that an in-depth analysis of a cross-section of media coverage at a particular point in time can reveal hitherto undetected cultural variations in how climate change is viewed and responded to. Importantly, it helps to make visible what Maarten Hajer (1993) aptly called discourse coalitions, that is, the kinds of stories that particular groups of social actors tell and that inform their everyday practices.

The remainder of this chapter is divided into four sections. The immediately following section combines a review of the relevant literature on climate cultures with the development of a conceptual framework for this study. This is followed by a detailed outline of the methodological design of this study. After that, the main empirical part describes and typifies the content of climate change debates in the media, distinguishing between statements made by members of the elite and views expressed by members of the general public (and with the latter often reacting to the former). Finally, we offer a discussion of the results and some concluding remarks.

The role of knowledge and action in climate cultures: conceptual foundations

Responsibility, efficacy, and everyday life: defining climate cultures

Although fundamental to public opinion and political (in)action regarding climate change, culture remains a marginal concept in climate change research. This is partly due to the ubiquity of cultural phenomena, including people's interactions with their natural environment, as well as the relative resistance of culture to conventional scientific definition and measurement. Consequently, media coverage and public debates concerning climate change tend to be "culturally blind", too, in addition to ignoring pressing social problems arising from a changing climate such as resource scarcity, poverty, and forced migration. This coincides with an almost complete disregard for people's everyday experiences and daily practices that characterize debates on environmental challenges more generally, and climate change, in particular. This reluctance to mind the mundane (Rau 2018) means that much climate change communication actually serves to disengage citizens (Fox and Rau 2017). This includes fear-inducing catastrophic and apocalyptic messages that incorporate a rather limited view of human agency and may thus prevent public climate action (Kundzewicz et al. 2020). Importantly for this study, such a perspective ignores lived experiences of responsible climate action and of individual and collective efficacy.

This said, there is some emerging literature that deals explicitly with linkages between culture and climate change. Thorsten Heimann, a leading contributor to research on climate cultures, attributes the concept to Claus Leggewie's research on climate change-related questions of "social responsibility, cultural memory and intercultural differences" (2009, 176). Similarly, Welzer et al. (2010) argue that climate-related social-scientific inquiries

> must consider the cultural practices and contexts of meaning that have caused climate change, thereby challenging human (...) sense-making and the philosophical consideration of aspects of justice and responsibility (...) as well as the knowledge-sociological analysis of collective interpretative patterns.
>
> (2010, 13)

Kari Norgaard's (2011) groundbreaking inquiry into the links between climate change, emotions, and everyday life has also quintessentially inspired this study. Her culturally sensitive sociological approach revealed the inherently social nature of *denial* that affects her respondents' reactions to impending climate change. Importantly, it shows how individualistic explanations of denial that are common in psychological studies fail to grasp collective decisions to ignore a particular threat (here: a Norwegian community's denial of climate change and its wide-ranging impacts on local livelihoods). Norgaard's (2019) work convincingly demonstrates the existence of different "cultures of denial" (see also Sutton and Norgaard 2013). More recent work by Norgaard and others highlights how public climate conversations largely omit social and cultural differences that influence people's views and practices regarding climate change, marginalizing forms of climate engagement that do not match scientific and societal expectations.

Building on these seminal contributions, we define climate cultures as dynamic variants of social organization that provide a framework for recognizing culturally relevant information regarding climate change and that are (re-)produced through climate-relevant everyday practices that reveal diverse forms of lived responsibility and actual experiences of (in)efficacy. The latter includes responses to more abstract attributions of responsibility and efficacy in official climate change discourses that potentially clash with people's everyday experiences. For example, the idea that individuals can contribute by consuming differently, that is, by buying green products (instead of consuming less) may fly in the face of those who are struggling to make ends meet. A concept of climate cultures that respects ordinary citizens' daily lives thus questions the excessive attention given to individuals' responsibility for "the right" (consumption) decisions that characterize many prominent approaches to climate action.

Responsibility and efficacy in knowledge and practice

Appealing to people's responsibility and demonstrating their capacity to act are central to effective climate action (e.g. Buschmann and Sulmowski 2018, 283, see also Vogt 2018). Responsibility attributions to different societal actors (individuals, politicians, the media, private sector, scientists) and expectations regarding these actors' capacity to act are therefore essential to the identification of climate cultures, including those of "knowledge elites" that dominate many public climate debates.

> Responsibility does not appear as an overarching, universal, and cross-temporally valid concept, instead, time and again, it is produced afresh and in different forms as a concrete and both historically and culturally situated, practice-specific phenomenon (Buschmann and Sulmowski 2018, 287).

Regarding the link between responsibility and efficacy, renowned social psychologist Albert Bandura states that "the voices for parochial interests are typically much stronger than those for collective responsibility"

(1995, 37) and that feelings of low efficacy can lead to hopelessness, apathy, and paralysis. Shared ideas, conventions, and interests and their translation into collective action can protect people against these negative feelings and experiences. This interpretation of efficacy as an inherently social phenomenon is central to this study's concept of climate cultures Focusing on the collective level when talking about climate cultures, can yield important clues as to why people (do not) act when confronted with the climate challenge.

Incidentally, recent reflections on denial and climate change have also pointed to notions of responsibility. Stoll-Kleemann and O'Riordan (2020, 1) observe a recent shift in public climate debates whereby "direct denial of anthropogenic climate change is replaced by a denial of responsibility for individual climate action". Processes of moral disengagement such as the diffusion of responsibility are central to this. Divergences also exist between expectations of efficacy and actual experiences of influence, as reflected in variations in discursive and material practices across different climate cultures discussed below.

A culturally sensitive analysis of people's responses to climate change that explicitly recognizes their inherently social nature serves to close at least some of the gaps left behind by conventional behavioral explanations of climate inaction. Three aspects deserve particular attention here: first, the reproduction of conceptual and methodological individualism; second, the persistence of information deficit thinking, and third, the omission of everyday practices and their links with both attributions and actual manifestations of responsibility and efficacy. The last point is particularly pertinent because divergences between abstract attributions and lived experiences concerning responsibility and efficacy appear to be central to variations in climate culture.

Moving beyond conceptual and methodological individualism

The limitations of prevailing individualistic models of human behavior, including studies of public understanding of climate change, have been well documented (Fox and Rau 2017; Shove 2010). Responding directly to these, this study moves beyond an exclusive focus on individuals and their (shared) knowledge.

> Knowledge is not what resides in a person's head or in books or in data banks. To know is to be capable of participating with the requisite knowledge competence in the complex web of relationships among people, material artefacts, and activities (Gherardi 2009, 517).

This means shifting the focus "away from the behaviour, action and motives of monological individuals", to view them as "members of groups and communities that constitute the context of their mundane activities" (Savolainen 2007, 120). According to Nerlich et al. (2010, 2), risks to humanity posed by climate change "are still for many largely 'virtual' risks rather than real ones", which encourages people to rely on pre-existing cultural repertoires or toolkits of solutions to everyday problems (Edmondson and Rau 2008; Swidler 1986), rendering climate change no longer an exclusively scientific but also and importantly a cultural issue (Nerlich et al. 2010, 2).

Challenging information deficit thinking

Much existing social-scientific literature on climate (in)action emphasizes information provision in changing unsustainable behavior (e.g. Blättel-Mink 2010, 30). This insinuates lay people lacking adequate knowledge to recognize the urgency of climate action. However, this type of "deficit thinking" has been subject to growing criticism, including for its neglect of cultural influences on climate (in)action (e.g. Suldovsky 2017). Importantly, deficit thinking reduces responses to climate change to cognitive processes, thereby ignoring the significance of affective, emotional, and bodily aspects of knowing, including people's engagement in everyday practices that shape their experiences of being in the world (Greene 2018; Davies et al. 2016). According to Lidskog et al. (2020, 118),

> for science to promote action, it is not sufficient that scientific advice is seen as competent and trustworthy. Such advice must also be perceived as meaningful and important, showing the need and urgency of taking action.

A growing body of work thus emphasizes the role of culture in constructing (climate) knowledge. For example, Heimann and Mallick (2016, 1) criticize that "factors to explain differences in perceiving and handling climate change besides shared knowledge remain blind spots". To respond to these blind spots, they develop an integrated knowledge concept that includes "shared cognitive and normative framings (e.g. shared problem framings for climate change, general values, beliefs, and identities) as well as shared practices at the level of action" (Heimann and Mallick 2016, 1).

Mirroring Heimann and Mallick's concerns, we adopt Olsson and Lloyd's notion of knowledge as "embodied information practices" (2017, 1) that recognizes the centrality of non-linguistic, experiential types of knowledge. Their work emphasizes the centrality of the body in knowledge acquisition and use because "information landscapes are not only shaped and represented socially and dialogically but also reflected corporeally" (Olsson and Lloyd 2017, 8). This reflects Richard White's observation that "(e)xamples of human knowledge of nature, gained through labor, are readily apparent if we look" (1996, 177). Building on these ideas, this study emphasizes the inherently collective and frequently tacit nature of knowing that is practiced within each climate culture.

The centrality of everyday life

With this comes an explicit recognition of the centrality of everyday life with its many routine practices as well as their relative resistance to efforts to transform them (Rau et al. 2020; Rau 2018; Sahakian and Wilhite 2014; Spurling et al. 2013). For example, eating habits and mobility practices have proven particularly difficult to change (Godin and Sahakian 2018; Heisserer and Rau 2017). This, in turn, is key to understanding the public's (lack of) engagement in climate action. Cultural norms and prescriptions are central to this resistance as they facilitate conserving a practice across multiple generations of practitioners (Shove and Walker 2010).

By acknowledging people's capacities to creatively solve problems in everyday life, for example by combining established routine practices to form new ones, a practice-centered perspective is uniquely suited to advance a view of human agency as socio-materially embedded (Rau 2018, 219).

Arguably, the apparent stickiness of many routine practices does not necessarily cause stagnation. Instead, people's insistence on doing things in a certain way can also help them to handle the unpredictability of everyday life. Therefore, a practice-centered perspective raises new ways of understanding people's climate (in)action and also opportunities for mobilization.

Regarding affective and emotional aspects of knowing and their manifestations in practice, this study recognizes shared forms of not knowing that shield people from unpleasant facts that challenge their way of life. Climate inaction may thus result from feeling intensely overwhelmed by information (cf. Lidskog et al. 2020). Observable expressions of socially organized denial represent a prime example of collective efforts toward not knowing (Stoll-Kleemann and O'Riordan 2020; Wallace-Wells 2020; Norgaard 2011). The more information people have, the more paralyzed, hopeless, and futile they might feel (Norgaard 2011). Consequently, people collectively "distance themselves from information because of norms of emotion, conversation, and attention, (…) [using] (…) an existing cultural repertoire of strategies in the process" (Norgaard 2011, 9). Evidence of such "distancing from information" (like media coverage of public debates) can uncover processes of socially organized denial and their adoption by different climate cultures.

Data and methodology

Media coverage of climate change provides particularly rich evidence of climate-cultural variations, a fact that remains seriously under-appreciated in much social-scientific work on culture and climate change. This study responds directly to this research gap by examining diverse media reports on climate change in Germany around the 2019 European elections, with a view to demonstrating which climate cultures (do not) feature in public debates on this important topic. A qualitative approach to data collection and analysis was chosen, with data deriving from four interrelated sources across TV, print, and social media.

First, debates on climate change in three leading German primetime political talk shows were considered: *Hart aber fair (hard but fair)* moderated by Frank Plasberg (25 March 2019), *Anne Will* (5 May 2019), and *Markus*

Lanz (27 June 2019). The format of these three talk shows is very similar, with prominent politicians and public figures discussing current affairs in a more or less confrontational way.[2] This said, *Markus Lanz* deviates somewhat from the other two in that this show seems a little less elitist due to the less formal demeanor of the talk show host and the wider selection of guests which often includes both prominent political figures as well as members of the public with a story to tell.

Second, we analyzed comments and discussions posted on two social media platforms (Facebook and Twitter) following social media postings by the producers/administrators (when possible) of the three talk shows. Examples of dialogue and exchange received particular attention. In connection to *Hart aber fair*, the analysis covered two threads on Facebook and Twitter that emerged in response to the administrator's introductions of the talk show guests Ulf Poschardt, editor-in-chief of the conservative *Welt* news group, and the then German Minister for the Environment Svenja Schulze. The editorial team of *Anne Will* does not maintain a Facebook account but two Twitter threads were analyzed. Regarding *Markus Lanz*, it was more difficult to find social media content, so the data was limited to one Twitter thread.

Third, we used online news articles from influential German news producers and leading weekly political magazines that covered, and commented on the three talk shows.

Fourth, we analyzed four YouTube Videos[3] by the two well-known Influencers *Rezo* (channel: *Rezo ja lol ey*) and *Mai Thi Nguyen-Kim* (channel: *maiLab*) on the topic of climate change.

The choice of sources reflects the intention to capture as broad a range of views and statements as possible (without being representative in the statistical sense of the term). Data collection took place from October to December 2019 and was carried out almost exclusively online (with the exception of a printed version of an article in *Der Spiegel* (23/2019) covering the posting of the Rezo video and its aftermath). The talk show data was transcribed and subsequently translated from German into English, complementing printed text and social media comments.

The wider context within which the media coverage emerged is also relevant. At the time of the 2019 European Election, the *Fridays for Future*-movement in Germany was gaining influence. There was a public call for increased climate protection and a push from numerous directions that politics should prioritize the issue. It ceased to be the exclusive topic of the Green Party, making its way into the political mainstream. A so-called climate cabinet was established two months prior to the election, with the aim of synthesizing legislation for mitigating climate change. Much of the data considered thus revolves around if and how climate change should be approached politically.

Using Kuckartz's (2012) typifying approach to qualitative content analysis, the collected data was examined to identify and subsequently compare statements regarding climate action and related issues of responsibility, efficacy, and knowledge. Given that there is no societal consensus regarding who should take the lead in promoting climate action in Germany, the responsibility attributed to, and efficacy expected of different societal actors received particular attention. The iterative analytical process combined deductive and inductive elements. An *a priori* interest in the role of different societal actors in climate change debates, the use of apocalyptic/catastrophic vocabulary, and the topic of individual responsibility informed the initial choice of methodology and data. The subsequent identification of different climate cultures happened inductively through a very close and repeated reading of the data.

Public debates about climate change: variable voices

This section presents the results, combining descriptive statements and direct quotes with interpretative efforts to ensure maximum readability, accessibility, and analytical transparency. The data was initially sorted into two categories: on the one hand, we detected a comparatively homogeneous set of rather visible climate cultures associated with political elites, prominent public figures, and well-known influencers. These elite climate cultures reflect ways of knowing, sense-making, and speaking associated with educated elites in German society, including choice of language and how scientific information is handled. What makes these climate cultures (including the subculture of young activists and influencers) "elite" is their exclusionary nature, given that only highly educated and media savvy members of the public can follow and actively participate in these discourses.

This contrasts with a less visible, heterogeneous group of climate cultures "from below" that comprise diverse views held by members of the general public. Here, various ways of understanding, arguing, and reasoning can be detected, many of which relate to everyday experiences of lived efficacy and responsibility. Some of these public views explicitly contradict or challenge elite practices and dominant ways of knowing while others accept their supremacy while offering potential alternatives. A subsequent fine-grained analysis of the data revealed four separate climate cultures – two linked to elite actors' public statements, and another two from the general public. Additionally, some displayed certain variations or shades that were found to comprise a subculture. In this study, a subculture represents a variant or shade of the cultural leanings associated with the parent culture. While internally the subculture strives to achieve some distinction from the parent culture, upon closer examination these deviations remain subtle enough to place them within the same general cultural category.

The first elite group of climate cultures finds expression in the three TV talk shows, online news portals, prestige print, and influential political magazines and, perhaps more surprisingly, climate-related YouTube clips by young influencers that received significant attention.[4] In contrast, the second group of climate cultures "from below" features very prominently across different social media, including comments sections linked to the aforementioned talk shows. In addition, some of these alternative climate (sub-)cultures appear in alternative media outlets such as print magazines focusing on green lifestyles and sustainability issues and political magazines that endorse anti-establishment views. However, these were not included in this study, due to their extensive range and diversity. Admittedly, some degree of overlap exists between some of the climate (sub-)cultures, for example regarding trust in expert opinions. However, we nevertheless decided to distinguish between the four climate cultures and their respective subcultures because of fundamental differences in key areas such as attributions of responsibility and expectations of efficacy vis-à-vis actual experiences of lived responsibility and efficacy.

Elite climate cultures

Two distinct elite climate cultures – individualist and collectivist – emerged from the analysis, with the second displaying a distinct subculture of young activists and influencers. Among these elite groupings, similarities include a shared language associated with official positions on climate change and action as well as more or less explicit acknowledgments that anthropogenic climate change exists and presents a serious challenge to humanity. At times, speakers from this elite category see themselves as well-informed and sufficiently competent to educate the public. For example, the arguments and terminology used by prominent influencers and YouTubers reveal their high educational status (Rezo holding a masters and Mai Thi Nguyen-Kim a doctoral degree) and their commitment to informing the public about climate change. Rezo's video clearly demonstrates his ability to speak the language of science (e.g. citing, summarizing, and synthesizing relevant studies). This is also evident when analyzing the open letter video signed by 90+ YouTubers. Here, the authors of the letter speak of "risk hierarchy", "scientific consensus", and being "discredited". All three elite climate (sub-)cultures feature some or all of the following arguments, many of which relate directly to aspects of responsibility and efficacy discussed in the second section of this chapter:

- Responsibility of political sphere versus individual responsibility.
- Efficacy of individual purchasing decisions.
- Social fairness of political decisions.
- References to the role of the private sector (responsibility and efficacy).
- Efficacy of scientific knowledge/technological innovations (perceived vs lived efficacy).
- Use of apocalyptic and catastrophic vocabulary (perceived vs lived efficacy).
- The questioning of the growth-dependent capitalist free-market system (perceived vs lived efficacy).

Elite with individualist tendencies

Current debates on climate protection in Germany frequently attribute responsibility for climate protection to the *individual* citizen-consumer, asking them to reduce unnecessary consumption to arrest climate change. Such views characterize this first variant of the elite category, coinciding with a more or less direct rejection of any legislative restrictions:

> In the past, actually just a couple of weeks ago, it was all about your own very personal (carbon) footprint that you leave our offspring with every light-pink T-Bone-Steak (...). This has now been literally chewed out, however, now it is about way more... Ulrich Reitz (journalist, FOCUS-online)

Here, Reitz's sarcasm indicates his rejection of lifestyle questions being made anything but his own business. Ulf Poschardt also vehemently stresses individual responsibility. Politicians who fall into this climate culture also tend to call for minimal political intervention to avert climate change. For example, they argue against a carbon tax, citing low public acceptance for such measures.

Interestingly, representatives of this climate culture ascribe considerable responsibility to private companies but, at the same time, believe that these already take sufficient responsibility given their exposure to other economic pressures such as profitability and competitiveness. In this context, the power and influence of individual consumers (*Verbrauchermacht*) is emphasized once again.

In contrast, experiences of lived efficacy do not correspond to this official discourse by members of this particular climate culture. Instead, representatives express little or no confidence in individual behavior change and self-restriction as effective means of climate action. Although social fairness is seen as essential to social stability, many members of this climate culture reject political efforts to redistribute wealth. Instead, trust is placed in the market to steer companies in the right direction, for example to develop technology to advance climate protection. Expert advice and scientific insights are seen as central to effective climate action. Additionally, rational decision-making is deeply appreciated, translating into high levels of skepticism concerning the role of emotions in climate discourse and action. An exception to this lies in Ulf Poschardt's explicit recognition of the centrality of emotions in consumption:

> a car is way more than an object that lets you travel from A to B. (...) I think the question must be how electro-mobility can also be emotionalized. And here I remain unconvinced because the electro-cars that I have driven (...) have no soul.

Nevertheless, the overall privileging of factual knowledge also puts responsibility on individuals to inform themselves to make the right decisions. This, in turn, deflects attention away from the responsibilities of more influential societal players. Lastly, catastrophic and apocalyptic statements are largely absent from this climate culture and only feature insofar as they are being criticized.

Elite with collectivist orientations

The most prominent difference between the individualist and this collectivist climate culture lies in their divergent attributions of responsibility:

> Robert Habeck (the then co-leader of the Green Party): (...) the point is that we live in serious political times and that our party is expected to carry an amount of responsibility like never before and we work extremely hard to do justice to this expectation. (...) And now we have the situation that almost every day another stone is being added to our backpack and then someone says, 'come on, run faster'!

The collectivist climate culture acknowledges the gravity of the climate crisis and the resulting need to act. Importantly, it ascribes much responsibility to political agents across the party-political spectrum and underlines the need for collective action. Criticism of current government inaction features prominently, an argument that is largely absent from the conservative climate culture described above. Members of this climate culture also caution against a narrow focus on individual behavior, questioning the impact of lifestyle changes. Instead, they focus on the political sphere as a main lever of change (lived efficacy). Another possible reason for redirecting attention away from individual consumption may lie in efforts to avoid or refute accusations of hypocrisy that are regularly leveled at elite advocates of climate action, most notably members of the Green Party and *Fridays for Future*-activists. Statements suggest that individual responsibility is overemphasized in public climate debates, not least because it stands in direct contrast to the limited influence individual consumption can actually have (lived efficacy). Some members of this climate culture nevertheless argue for a shift in individual consumption habits, recognizing the shared moral imperative to do so (hence the label "collectivist"). Still, the impact of the private sector is considered to be high (lived efficacy), mirroring some of the statements made by members of the more individualist climate culture.

Concerning social fairness, this climate culture argues for sharing the burden of climate action much more equally than is currently the case. Politics is deemed to be largely responsible for organizing this burden sharing, including through financial redistribution. This emphasis on the (interventionist) role of politics stands in stark contrast to the market-focused, laissez-faire perspective endorsed by members of the individualist climate culture.

With respect to knowledge and technological innovation, some parallels however exist between the two elite cultures: information deficit discourses, a prioritization of science and faith in expert-knowledge loom large. However, apocalyptic stories are present to some extent because they reflect the idea that strong negative emotions (e.g. fear) mobilize people.

Lastly, many members of this collectivist climate culture blame contemporary production and consumption infrastructures, politics, and governance for accelerating climate change. However, not everybody expects solutions to emerge from a radical systemic change. According to Jan Grossarth (journalist, SZ),

> differentiating and balancing, out of responsibility, whilst aware of the complexity, that'd actually be the more bourgeois-conservative approach. (...) What would a family business be without a lively environment? (...) in the long-term, climate protection in fact supports the conservation of the market-based system.

As this shows, some representatives of this culture believe in the effective contribution of the market economy to successful climate protection.

Collectivist subculture: emerging elite discourses of activists and influencers

The discourses of young German activists and influencers represent a subculture of the collectivist elite climate culture, as the discursive practices of its members closely resemble those of the parent culture. They stress the importance of information provision, coupled with a firm belief in the effectiveness of scientific reasoning. Representatives also frequently deploy scientific arguments, techniques, and terminology to promote climate protection. For example, YouTubers Rezo and maiLab regularly refer to scientific studies and prominent climate scientists and campaigners. Here, for example, Rezo refutes the commonly raised argument that switching to renewables would destroy too many German jobs:

> Well, maybe (one could say) our efforts are being thwarted by other countries? Maybe we want to do a lot... What? Okay, other countries are starting initiatives (...) and we are thwarting that and are not joining in? Huh. Right, ok, CDU maybe is thwarting the fight against global warming, but hey, maybe they have a good reason. So economically, this is unbearable? Coal is huge in Germany... What? Only 20.000 jobs in the whole of the coal sector?

They also cite particular scientific findings to rebut skewed and misleading arguments used by climate sceptics to deflect responsibility (e.g. accusation of eco-hypocrisy aimed at politicians, ad hominem arguments). While internalizing individuals' moral obligation and responsibility to contribute to climate protection, members of this subculture also question the efficacy of the individual, ascribing responsibility and efficacy to the political sphere instead. At the same time, they doubt the ability of current political leaders to legislate for effective climate protection. The willingness of private-sector actors to engage in serious climate protection and to do so voluntarily is questioned too, revealing a discrepancy between attested responsibility and efficacy. For example, some YouTubers view private businesses as very powerful actors that will always put their interests first, unless laws and regulations force them to act in the interest of the climate.

Communicative mechanisms for promoting immediate climate action are a key interest of members of this subculture. Here, the explicit use of colloquial language serves the purpose of motivating its primary target audience (young people) to accept their responsibility and act accordingly. Examples include Rezo's use of the term "mate" (*Diggi*) to address his viewers or Luisa Neubauer's use of "most-massively" (*massivst*) to accentuate the scale and urgency of the climate change problem. maiLab also uses youth language but to a lesser extent as she directs her videos at young people interested in science. Moreover, members of this subculture highlight the central role of emotional triggers in mobilizing different societal actors to accept their responsibility for climate protection and to act accordingly, which is what sets them apart from the other two elite climate cultures discussed previously. It is for example regularly resorted to catastrophic language to mobilize audiences to take immediate climate action. As Rezo states,

> the predicted mass extinction (...) can then also not be stopped (...). But we get our own little goodie-bag: there will actually be more diseases for us, mainly more infectious diseases and allergies. Jippieh!

Yet, the "system question" is hardly ever raised within this elite subculture. Overall, the speakers of these elite climate cultures are unified by their implicit and explicit acceptance of official narratives regarding the societal imperative to "do something about climate change", albeit with varying degrees of urgency.

Climate cultures "from below"

Only by analyzing the public's reactions to the ideas featuring among elite climate cultures, Germany's climate-cultural diversity becomes fully apparent. Climate cultures "from below" appear to be significantly more heterogeneous and eclectic than those in the elite group, which is not yet adequately recognized in climate-related research and policy. This invisibility of public climate-related views and practices can be partly attributed to them rarely having access to the same kinds of elite platforms (e.g. political talk shows, prestige print media). Two distinct climate cultures (action and inaction) were identified, with the latter displaying three subcultures (inefficacy, skepticism, and denial/anti-[youth]activism), each of which displays its own discursive patterns and logics. Perhaps most surprisingly was the surfacing of a broad range of manifestations of climate change skepticism and even denial. This was unexpected insofar as German society is not usually described in the literature as particularly climate-skeptical (cf. Walter et al. 2018; Tranter and Booth 2015; Grundmann 2007). What also stands out is the polarization of the public debate between radical pro-climate discourses on the one hand, and skeptical and denialist discourses on the other.

Pro-climate action culture

This first climate culture "from below" is characterized by its close links to green values that are ultimately also mirrored in political affiliation. Opinions among its members tend to revolve around the following points:

- Debates about the individual holding responsibility and "quiet" acceptance of some form of individual responsibility.
- Rejection of conspicuous consumption.
- References to the "climate consensus".
- Strong support for, and celebration of, the electoral success of the Green Party in the 2019 elections.

Importantly, its representatives embrace the idea that people should "practice what they preach", signaling efforts to rebut eco-hypocrisy arguments.

> What nonsense! It has never been easier for the individual to avoid CO_2: heaps of cheap meat, cheap flights, cruises, skiing holidays 'vacations'? All are ways to practice what you preach and fast CO_2. But this can only be done by practicing restraint. (Christian Wirth)

This coincides with moralist appeals to be consistent and authentic and to set a good example. Furthermore, some expressions are quite alarmist:

> If we don't change radically, through our actions we will be the killers of our children. This is a fact. No matter what any capitalist marionettes say, this will be fact. (Unbequeme Wahrheit)

At the same time (and in contrast to the collectivist elite culture), there is little evidence of this climate culture recognizing the many real practical obstacles that prevent citizens from leading their everyday lives in a climate-friendly way. Responding to Kevin Kühnert [prominent member of the Social Democrats] addressing this issue in the *Anne Will* talk show, some members of this climate culture raise the "system question" and related aspects of climate justice:

a politics relying on annual growth of 2% is going to miss any climate targets in this world. A politics that enters into economic contracts with countries that will obviously be exploited through them does not deserve the name politics!

(G.D.S.)

They nevertheless denounce the apparent inefficacy of the current government with regard to climate protection. Scientific experts are again well respected by members of this climate culture, and there is some significant overlap with the collectivist elite climate (sub-) cultures concerning responsibility attributions to different societal actors.

Inaction climate culture(s)

By contrast, this second climate culture "from below" is characterized by its (at times fierce) rejection of current public calls for more climate action. As reasons for this rejection range from experiences of inefficacy to outright climate change denial, statements have been assorted into three subcultures.

Subculture I: sense of inefficacy. This first subculture is distinguished by a lack of trust in the effectiveness of official climate protection measures. Members remain largely unconvinced that society and individuals can afford wide-ranging climate protection, pointing to the high cost of measures such as the promotion of electric vehicles or the purchasing of organic food.

why am I, as individual consumer, supposed to mind that the products offered to me increase business profits and, at the same time, ruin the environment? Why do you ask me to take responsibility for the damage they do? (Petra Meier)

Issues discussed include the limited efficacy of the individual in their private sphere and the inherent unsustainability in work-life and industry that seem of much higher impact.

as long as I HAVE TO cover each palette at work with X meters of cling-film for load securing but am expected to pay for a plastic bag because of the environment, I'm gonna kill myself laughing. (Christian Beetz)

In addition to questioning the effectiveness of proposed climate protection measures, there is also doubt whether they could be reconciled with social fairness. Due to its global nature, national-level efforts to avert climate change are deemed insufficient (directly opposing Rezo's statement above). Finally (and contrasting its pro-action counterpart), this subculture is unconvinced that climate protection can feasibly and practically be integrated into their everyday realities (lived efficacy), even when they acknowledge that they would have at least some responsibility to act.

Subculture II: skepticism. There is, however, ample evidence of more radical objections to current calls for climate protection across the data studied, yielding a second inaction subculture. Its members tend to view climate change and related calls for action as *scaremongering, hysteria,* or a *ridiculous hype.*

> I couldn't find the facts mentioned by Greta. Which are they? I increasingly understand the climate sceptics! The changing climate cannot be related to the increase in CO_2-concentration. I'm technically trained. (Ludwig)

Discourses regarding individuals' responsibility are not featured. However, there are regular references to morality and efficacy, for example when members talk about double standards and critique costly measures that to them occur on "secondary battlefields" (i.e. in areas of everyday life that members of this climate culture do not consider to be very important).

> all this is nonsense. Do they really believe what they ask? Where is all the electricity going to come from if there are only e-cars? Wind turbines? Of course! Wants everybody in their front yard. The electricity also has to be transported. (...) What about recycling the batteries? Nobody talks about that. All nicely hidden from view. All of this is a farce. (Andreas Vogler)

Here, politicians are generally deemed untrustworthy and criticized for placing the burden of crisis management largely on the shoulders of the public.

Subculture III: denial. Even more radical yet, in several social media threats the truth content of current climate science was questioned, complemented by more or less overt discourses of climate change denial. For example, in some cases, climate change was treated as a conspiracy, led by the rapidly growing climate movement. Like in the two previous inaction subcultures, establishment politicians and institutions (for example the IPCC) are largely deemed untrustworthy, albeit more harshly this time.

> hello Mr. Lanz, you haven't replied yet. So I ask again: when are you inviting the dual leadership of the AFD? I only ever see the Greens in all talk shows, is public TV practicing electoral influencing? (Oberschlesier)

This excerpt exemplifies a clear link between rightist/nationalist political orientations and hostility toward environmentalism, in general, and climate change policy, in particular. This position regularly coincides with nativist and anti-immigrant views, a hostility toward all elites (scientific, academic, political, cultural, etc.) and strong support for individual choice and autonomy. Members of this subculture also reject any attribution of responsibility to themselves for the living conditions of future generations, for example by discrediting climate activism among young(er) people through practices of ridicule and emphasizing an alleged lack of life-experience and maturity. Climate activists' calls to assist those most affected by climate change (now and in the future and here and elsewhere) are interpreted as attempts to further limit people's freedom.

Discussion and conclusions

Climate-related debates that occur across different media formats are far from consensual and reveal variable voices that harbor the potential for serious societal conflict. For example, variations exist regarding the significance and urgency ascribed to the challenge of climate change and measures to address it. Importantly, a significant gap exists between climate-related arguments and debates that members of elite climate cultures engage in, and those that can be assigned to climate cultures "from below". Here, it is possible to identify culturally distinct notions of responsibility and self-efficacy and divergent ideas around what counts as acceptable knowledge. This shows that the ways in which people talk about climate change link more or less directly to cultural norms and conventions that contribute to the social regulation of everyday life, and that guide and shape people's engagement in climate-relevant practices across domains such as food and mobility. Our analysis revealed a significant (and potentially widening) gap between elite climate cultures and those attributed to the general public, pointing toward high levels of climate-cultural diversity in Germany. In some cases, there appears to be a complete disconnect between those who debate climate change and climate action on mainstream media and those who use social media channels to express their views. Moreover, the nature and content of many elite contributions to the debate appear to be of limited relevance to members of the public who comment on climate issues through social media channels, including those attached to mainstream media such as TV talk shows. These observable climate-cultural divergences are likely to be responsible for the slow progress in relation to climate change mitigation and adaptation, at least to some degree, a fact that remains under-appreciated in scientific and public debates on climate action.

Discursive variations that revealed themselves through our in-depth analysis of media coverage of climate change topics around the 2019 European elections also highlight the centrality of notions of responsibility and self-efficacy for understanding clashes between climate cultures. Here, a decoupling of responsibility and efficacy is clearly discernible across a number of climate cultures. For example, individual consumers are routinely blamed for not doing enough to protect the climate, including by climate activists. This places the burden of responsibility on the shoulders of those who are least able to act. In contrast, powerful societal actors like politicians and business leaders are portrayed as being too self-interested to act decisively on climate change, despite their actual capacity to do so. This (perceived) inefficacy and irresponsibility of established actors is utilized by other political and civil society actors to challenge dominant scientific and official climate cultures and to advance counter-arguments ranging from a radically green, pro-climate agenda to variants of climate change denial.

Climate cultures that have formed both within and outside elite circles also incorporate a broad range of emotive aspects that relate very closely to notions of responsibility, efficacy, and knowledge. Elite discourses frequently emphasize the centrality of rationality and scientific knowledge in climate protection. At the same time, they reveal beliefs of an inherent information deficit among members of the public that targeted climate education can help to overcome. This emphasis on the cognitive dimensions of knowledge contrasts with climate debates "from below" that are interspersed with references to embodied knowledge, everyday practices, and emotions, lending support to the use of emotional appeals in climate action initiatives. The necessity to link both cognitive and emotional aspects of knowing in the context of climate debates and action thus emerges as a key challenge to those interested in advancing climate protection.

To conclude, the persistent (discursive) exclusion of certain climate cultures from mainstream media coverage serves to eclipse the existence of climate-cultural diversity in Germany, a fact that becomes particularly apparent when attention is shifted to social media outlets. Here, a wide range of different and at times conflicting climate cultures becomes apparent. Future efforts to progress climate protection in Germany (and elsewhere) need to take seriously this climate-cultural diversity, including its inherent conflict potential. We argue that an explicit recognition of climate cultures that emerge "from below", including those described in this chapter can help to overcome the persistent disengagement of large parts of the public from elite climate debates and cultures and related political and practical projects.

Notes

1 All translations were carried out by the first author.
2 The title of the talk show 'Hard but fair' in particular suggests a rather combative approach.
3 The maiLab videos were not published around the same time as the talk shows (15.09.19 + 22.03.18).
4 Clicks to date (04.06.2020). Rezo, Die Zerstörung der CDU: 17.347.533; maiLab, Klimawandel: Das ist jetzt zu tun!: 866.947; maiLab, Die Klimawandel-Therapie: 88.652; Ein Statement von 90+ Youtubern: 4.395.518.

References

Adams, John. 2007. "Risk Management: It's Not Rocket Science – It's Much More Complicated". *Public Risk Forum*. http://www2.sa.unibo.it/docenti/emanuele.padovani/PublicRiskForum_2_07.pdf

Backes, Laura, Tobias Becker, Lothar Gorris, Judith Horchert, Anna-Lena Jaensch, and Alexander Kühn. 2019. "Kinder der Apokalypse" [Children of the Apocalypse]. *Der Spiegel*, 20 May, 12–21. www.spiegel.de/politik/greta-thunberg-rezo-und-fridays-for-future-der-protest-den-wir-verdienen-a-00000000-0002-0001-0000-000164179769

Bandura, Albert. 1995. *Self-Efficacy in Changing Societies*. Cambridge (United Kingdom): Cambridge University Press.

Blättel-Mink, Birgit. 2010. "Konsum und Nachhaltigkeit – ein Widerspruch? Wie soziokulturelle Milieus Lebensstil und Konsumverhalten bestimmen" [Consumption and Sustainability – A Contradiction? How Sociocultural Milieus Determine Lifestyle and Consumer Behavior]. *Forschung Frankfurt: Wissenschaftsmagazin der Goethe-Universität* 28 (3): 26–30.

Buschmann, Nikolaus, and Jędrzej Sulmowski. 2018. "Von 'Verantwortung' zu 'doing Verantwortung': Subjektivierungstheoretische Aspekte nachhaltigkeitsbezogener Responsibilisierung" [From 'Responsibility' to 'Doing Responsibility': Subjectivation-Theoretical Aspects of Sustainability-Related Responsibilization]. In *Reflexive Responsibilisierung. Verantwortung für nachhaltige Entwicklung* [Reflexive Responsibilization. Responsibility for Sustainable Development], edited by Anna Henkel, Nico Lüdtke, Nikolaus Buschmann, and Lars Hochmann, 281–229. Bielefeld (Germany): transcript.

Edmondson, Ricca, and Henrike Rau, eds. 2008. *Environmental Argument and Cultural Difference: Locations, Fractures and Deliberations*. Bern (Switzerland): Peter Lang.

Davidson, Debra J. 2018. "Rethinking Adaptation". *Nature and Culture* 13 (3): 378–402.

Davies, Anna, Frances Fahy, and Henrike Rau, eds. 2016. *Challenging Consumption: Pathways to a More Sustainable Future*. London (United Kingdom): Routledge.

Fox, Emmet, and Henrike Rau. 2017. "Disengaging Citizens? Climate Change Communication and Public Receptivity". *Irish Political Studies* 32 (2): 224–246.

Gherardi, Silvia. 2009. "Introduction: The Critical Power of the Practice Lens". *Management Learning* 40 (2): 115–128.

Godin, Laurence, and Marlyne Sahakian. 2018. "Cutting Through Conflicting Prescriptions: How Guidelines Inform 'Healthy and Sustainable' Diets in Switzerland". *Appetite* 130: 123–133.

Greene, Mary. 2018. "Socio-Technical Transitions and Dynamics in Everyday Consumption Practice". *Global Environmental Change* 52: 1–9.

Grundmann, Reiner. 2007. "Climate Change and Knowledge Politics". *Environmental Politics* 16 (3): 414–432.

Hajer, Maarten. 1993. "Discourse Coalitions and the Institutionalization of Practice: The Case of Acid Rain in Great Britain". In *The Argumentative Turn in Policy Analysis and Planning*, edited by Frank Fischer and John Forester, 43–76. Durham and London (United Kingdom): Duke University Press.

Heidbrink, Ludger. 2003. *Kritik der Verantwortung: Zu den Grenzen verantwortlichen Handelns in komplexen Kontexten* [Critique of Responsibility: On the Limits of Responsible Action in Complex Contexts]. Weilerswist (Germany): Velbrück.

Heimann, Thorsten. 2017. *Klimakulturen und Raum: Umgangsweisen mit Klimawandel an europäischen Küsten* [Climate Cultures and Space: Coping with Climate Change on European Coasts]. Wiesbaden (Germany): Springer VS.

Heimann, Thorsten, and Bishawjit Mallick. 2016. "Understanding Climate Adaptation Cultures in Global Context: Proposal for an Explanatory Framework". *Climate* 4 (4): 1–12.

Heisserer, Barbara, and Henrike Rau. 2017. "Capturing the Consumption of Distance? A Practice-Theoretical Investigation of Everyday Travel". *Journal of Consumer Culture* 17 (3): 579–599.

Hulme, Mike. 2017. *Weathered: Cultures of Climate*. London (United Kingdom): Sage Publications.

Kuckartz, Udo. 2012. *Qualitative Inhaltsanalyse: Methoden, Praxis, Computerunterstützung [Qualitative Content Analysis: Methods, Practice, Computer Support]*. Weinheim (Germany): Beltz-Juventa.

Kundzewicz, Zbigniew W., Piotr Matczak, Ilona M. Otto, and Philipp E. Otto. 2020. "From 'Atmosfear' to Climate Action". *Environmental Science & Policy* 105: 75–83.

Leggewie, Claus, Harald Welzer, and Ludger Heidbrink. 2009. "Klimakultur – ein transdisziplinärer Projektverbund". In *Zwei Grad. Das Wetter, der Mensch und sein Klima* [Two Degrees. The Weather, Man and Climate], edited by Petra Lutz, and Lutz Macho, 176–182. Göttingen (Germany): Wallstein.

Lidskog, Rolf, Monika Berg, Karin M. Gustafsson, and Erik Löfmarck. 2020. "Cold Science Meets Hot Weather: Environmental Threats, Emotional Messages and Scientific Storytelling". *Media and Communication* 8 (1): 118–128.

Maranta, Alessandro, Michael Guggenheim, Priska Gisler, and Christian Pohl. 2003. "The Reality of Experts and the Imagined Lay Person". *Acta Sociologica* 46 (2): 150–165.

Nerlich, Brigitte, Nelya Koteyko, and Brian Brown. 2010. "Theory and Language of Climate Change Communication". *WIREs Climate Change* 1 (1): 97–110.

Neverla, Irene, Monika Taddicken, Ines Lörcher, and Imke Hoppe, eds. 2018. *Klimawandel im Kopf: Studien zur Wirkung, Aneignung und Online-Kommunikation* [Climate Change in the Mind: Studies on Impact, Appropriation and Online Communication]. Wiesbaden (Germany): Springer VS.

Norgaard, Kari M. 2019. "Making Sense of the Spectrum of Climate Denial". *Critical Policy Studies* 13 (4): 437–441.

Norgaard, Kari M. 2011. *Living in Denial: Climate Change, Emotions, and Everyday Life*. Cambridge, MA, and London (United Kingdom): MIT Press.

Olsson, Michael, and Annemaree Lloyd-Zantiotis. 2017. "Being in Place: Embodied Information Practices". *Information Research: An International Electronic Journal* 22 (1). https://discovery.ucl.ac.uk/id/eprint/10062466

Rau, Henrike. 2018. "Minding the Mundane: Everyday Practices as Central Pillar of Sustainability Thinking and Research". In *Environment and Society*, edited by Magnus Boström and Debra J. Davidson, 207–232. Cham (Switzerland): Springer International Publishing.

Rau, Henrike, Paul Moran, Richard Manton, and Jamie Goggins. 2020. "Changing Energy Cultures? Household Energy Use before and after a Building Energy Efficiency Retrofit". *Sustainable Cities and Society* 54: 1–13.

Sahakian, Marlyne, and Harold Wilhite. 2014. "Making Practice Theory Practicable: Towards More Sustainable Forms of Consumption". *Journal of Consumer Culture* 14 (1): 25–44.

Savolainen, Reijo. 2007. "Information Behavior and Information Practice: Reviewing the 'Umbrella Concepts' of Information-Seeking Studies". *The Library Quarterly* 77 (2): 109–132.

Schäfer, Mike S. 2012. "Online Communication on Climate Change and Climate Politics: A Literature Review". *WIREs Climate Change* 3 (6): 527–543.

Schäfer, Mike S., and Heinz Bonfadelli. 2017. "Umwelt- und Klimawandelkommunikation" [Environmental and Climate Change Communication]. In *Forschungsfeld Wissenschaftskommunikation* [Research Field Science Communication], edited by Heinz Bonfadelli, Birte Fähnrich, Corinna Lüthje, Jutta Milde, Markus Rhomberg, and Mike S. Schäfer, 315–338. Wiesbaden (Germany): Springer Fachmedien.

Schmidt, Andreas, Ana Ivanova, and Mike S. Schäfer. 2013. "Media Attention for Climate Change around the World: A Comparative Analysis of Newspaper Coverage in 27 Countries". *Global Environmental Change* 23 (5): 1233–1248.

Shove, Elizabeth. 2010. "Beyond the ABC: Climate Change Policy and Theories of Social Change". *Environment and Planning A: Economy and Space* 42 (6): 1273–1285.

Shove, Elizabeth, and Gordon Walker. 2010. "Governing Transitions in the Sustainability of Everyday Life". *Research Policy* 39: 471–76.

Spurling, Nicola, Andrew McMeekin, Elizabeth Shove, Dale Southerton, and Daniel Welch. 2013. "Interventions in Practice: Re-Framing Policy Approaches to Consumer Behaviour". *Sustainable Practices Research Group*, Manchester. https://eprints.lancs.ac.uk/id/eprint/85608/

Stoll-Kleemann, Susanne, and Tim O'Riordan. 2020. "Revisiting the Psychology of Denial Concerning Low-Carbon Behaviors: From Moral Disengagement to Generating Social Change". *Sustainability* 12 (3): 935.

Suldovsky, Brianne. 2017. "The Information Deficit Model and Climate Change Communication". In *Oxford Research Encyclopedia of Climate Science*, edited by Matthew C. Nisbet, Shirley S. Ho, and Ezra Markowitz et al. Oxford (United Kingdom): Oxford University Press.

Sutton, Barbara, and Kari M. Norgaard. 2013. "Cultures of Denial: Avoiding Knowledge of State Violations of Human Rights in Argentina and the United States". *Sociological Forum* 28 (3): 495–524.

Swidler, Ann. 1986. "Culture in Action: Symbols and Strategies". *American Sociological Review* 51 (2): 273.

Tranter, Bruce, and Kate Booth. 2015. "Scepticism in a Changing Climate: A Cross-National Study". *Global Environmental Change* 33: 154–164.

Vogt, Markus. 2018. "Theological Perspectives in the Ethical Debate about Climate Change". In *Climate Change and Cultural Transition in Europe*, edited by Claus Leggewie and Franz Mauelshagen, 60–82. Leiden (The Netherlands): Brill.

Walker, Gordon, Noel Cass, Kate Burningham, and Julie Barnett. 2010. "Renewable Energy and Sociotechnical Change: Imagined Subjectivities of 'the Public' and Their Implications". *Environment and Planning A: Economy and Space* 42 (4): 931–947.

Wallace-Wells, David. 2019. *The Uninhabitable Earth: Life after Warming*. New York: Tim Duggan Books.

Walter, Stefanie, Michael Brüggemann, and Sven Engesser. 2018. "Echo Chambers of Denial: Explaining User Comments on Climate Change". *Environmental Communication* 12 (2): 204–217.

Welzer, Harald, Hans-Georg Soeffner, and Dana Giesecke, eds. 2010. *KlimaKulturen: Soziale Wirklichkeiten im Klimawandel* [Climate Cultures: Social Realities in Climate Change]. Frankfurt am Main (Germany) and New York: Campus.

White, Richard. 1996. "'Are You an Environmentalist or Do You Work for a Living?' Work and Nature". In *Uncommon Ground: Rethinking the Human Place in Nature*, edited by William Conon, 171–185. New York and London (United Kingdom): W.W. Norton.

4 Unusual Suspects

Urban Change Agents for Climate Change Mitigation in Germany[1]

Miriam Schad and Bernd Sommer

Introduction

"Change agents" or "pioneers of change" are central players in transformation processes toward more sustainable relations between human and nature (WBGU 2011; Kristof 2010, 2020, 2021; Leggewie and Welzer 2010). Well-known examples are citizen cooperatives or local sustainability initiatives in the context of the "Transition Town" movement (Hopkins 2013). Against the backdrop of policy blockades, civil society actors working bottom-up for climate change mitigation and sustainability goals have gained additional attention. The concept of "change agent" originates in diffusion research (Rogers 1995) and change management (Beckhard 1969). In past transfers of the concept to interdisciplinary sustainability research, two dominant uses can be identified: first, following Rogers (1995), intermediary actors who facilitate the introduction of renewable energies are described as "change agents" (Mautz et al. 2008). Second, the concept is also used for prominent experts from the field of environmental protection, such as department heads of companies and public authorities, politicians, or outstanding representatives of non-governmental organizations, who advance climate and sustainability goals (see Clar and Steurer 2019a, 2019b; Kristof 2010).

However, if transformation into a more sustainable society requires cultural change (Leggewie and Mauelshagen 2018; Hoffman 2010; Welzer et al. 2010) and a broad participation of the population (WBGU 2011), the limited application of the concept to these two groups of people seems insufficient. Moreover, the discussion of the concept of "change agent" in the sustainability debate (Kristof 2010, 2020, 2021) rarely draws on any findings of sociological research and theory. For instance, in their study, Bliesner et al. (2013) attempt to determine the target profile of a change agent for sustainability. In this context, the understanding of change agents is limited to actors in a professional context or within companies and other organizations, while a comprehensive catalogue of personal and social competences such as healthy psychological constitution, sympathetic and winning personality, or social sensitivity is identified as the fundamental feature of the target profile.

DOI: 10.4324/9781003307006-8

Otherwise, change agents often appear as heroic individuals, which usually leaves the larger social contexts, as well as the necessary individual dispositions to set change processes in motion, underexamined. Our article takes up these two *desiderata* and explores the following three questions:

1 Which social and habitual conditions enable local actors to act as change agents toward sustainability?
2 Are local change agents other than typical pro-environmental actors open to addressing sustainability and climate protection issues?
3 What role do local change agents play for the formation of climate cultures on the local level?

To answer these questions, we draw on results of the research project "Knowledge Base for Individual Action: Change Agents for Climate Protection", which will be described below, after a brief section on theoretical foundations. In our analysis, we explore the social and habitual characteristics of change agents and their attitudes toward sustainability based on a qualitative field study in the German city of Cologne. We use Pierre Bourdieu's practice-theoretical approach to compare a disadvantaged neighborhood of Cologne with a privileged one. The central findings of our study as well as their relevance for the formation of local climate cultures are summarized and evaluated in the discussion section of this paper.

Theoretical foundations

In research on the diffusion of "green" innovations and the transformation to a climate-friendly society, Everett M. Rogers' standard work *Diffusion of Innovations*, first published in 1962, is still attracting attention (see Fichter and Clausen 2013; WBGU 2011). Based on a large number of individual empirical studies, Rogers (1995) describes the nature of innovations, their communication channels, and the temporal and (social) systemic dimension of diffusion processes. According to Rogers, the speed of adoption of innovations, and thus the diffusion process itself, depends, among other things, on early adopters, who are at the beginning of the diffusion process, opinion leaders, who disseminate relevant information about innovations, and so-called "change agents". Rogers defines change agents as individuals who influence the innovation-related decisions of other actors, such as consultants or sales representatives. Change agents thus assume a central mediating role in the diffusion process.

Rogers conceptualizes innovation relatively broadly as "an idea, practice, or object perceived as new by an individual or other unit of adoption" (1995, 35). Nonetheless, almost all examples on which he bases his diffusion theory are technological innovations. For Rogers, the social characteristics of diffusion agents also refer primarily to actors in the diffusion of technological innovations. He broadly states that early adopters, opinion leaders, and change agents have a higher level of education and social status, as well as a "cosmopolitan attitude", by which he means an openness to innovations and a willingness to take risks (Rogers 1995, 282 ff.). Apart from this, personal characteristics of diffusion agents, such as values or dispositions, remain largely unexamined. The substantive goals of technical innovations, or even their environmental impact, are only implicitly addressed at best (Fichter and Clausen 2013). In other words, Rogers' concept of "change agents", just like his entire diffusion model, do not take into account the specifics of transformation toward sustainability. Like this, the concept can also be applied, for example, to the spread of technologies that are highly ambivalent from an environmental protection perspective, such as hydraulic fracturing (fracking) for the extraction of shale gas.

To date, the adoption of the concept of "change agents" in interdisciplinary sustainability research has been somewhat undertheorized and largely uncritical. However, despite the undisputed role that is played, for example, by the diffusion of renewable energies, societal transformation according to the guiding principle of sustainability cannot be reduced to the diffusion of new technologies (Zell-Ziegler et al. 2021; Böcker et al. 2020; WBGU 2011; Hoffman 2010). In the following, by drawing on Pierre Bourdieu's theory of social practice, we offer a sociological foundation of the concept of "change agents" for interdisciplinary sustainability research. We also offer a substantive clarification of the concept based on the findings of a qualitative study on civic engagement in two neighborhoods of the German city of Cologne.

According to Bourdieu (1977, 1984), endowment with different types of capital determines an individual's position in social space and thus also their possibilities of making an impact on society. Bourdieu distinguishes between economic, social, cultural, and symbolic forms of capital, with economic capital playing a dominant role. By this, he means any kind of possession that is directly convertible into money. Modern social theory, as elaborated by Bourdieu (1984), Elias (2001), or Giddens (1986), also assumes that, in addition to social structures, there are subjectivized structures that guide social action. Following Bourdieu and Elias, these are referred to as "habitus". An investigation of the predispositions of actors toward becoming local change agents must therefore not be limited to material-social structures, yet must also take into account the mental dispositions, orientations, and value systems of the actors. Below, we will pay particular attention to these dimensions of "habitus", in addition to social characteristics that indicate various forms of capital. In doing so, we link to the discussion on climate cultures (Heimann 2019; Welzer et al. 2010), which raises the question of how different climate change related cultures emerge, socio-spatially (which also includes regions or cities), as well what kind of characteristics more climate-friendly, as well as more opposing cultures, have. Our research focuses on the emergence and formation of climate change mitigation practices in the context of different socio-economic surroundings, resources, and varying networks in two different neighbor-hoods. In this context, we do not assume a "single" culture in the respective district. Instead, we expect that socio-spatial orders of knowledge emerge through the relational networks of change agents.

Methods and data

The empirical research project was carried out in 2011 by the Institute for Advanced Study in the Humanities (KWI Essen, Germany) in cooperation with the KATALYSE Institute for Applied Environmental Research and financially supported by KlimaKreis Köln (a funding agency for climate change mitigation efforts; initiated by the University of Applied Sciences Cologne and the local energy provider RheinEnergie). Our research was part of a larger study titled "Dialogue Cologne Climate Change: A Green Master Plan for the City".

We proceeded as follows to collect the empirical materials: we first selected one disadvantaged and one privileged urban district in the city of Cologne. The sociological distinction between privileged and disadvantaged urban neighborhoods is based on processes of spatial segregation: people with similar socio-economic characteristics are disproportionately represented in certain districts and neighborhoods (Lloyd et al. 2014). The neighbor-hoods we studied differed markedly in terms of household size, unemploy-ment rate, and percentage of welfare recipients. In both neighborhoods, we identified people who were actively involved in local change processes, but who had no explicit connection to climate protection and sustainability is-sues. In doing so, we assumed that actors can drive social change processes "from below" or "bottom-up". In an explorative qualitative study, twelve participants were interviewed between April and November 2011. The guideline used contained questions on the topics of "commitment" (such as "Please tell us what you do as part of your work as an association chair-man?"), "local network" (such as "who do you work with locally and what does the cooperation look like?"), and "neighborhood" (such as "what do you like and what do you not like about your neighborhood?").

The topics of "ecological sustainability" and "climate protection" were not explicitly addressed in our interviews, in order to avoid socially desira-ble response bias. Instead, indications of a general openness to these con-cerns were to be gathered indirectly through questions about a "good life" and intergenerational justice. The related questions were "What constitutes a 'good life' for you? How would you like to live?" and "Do you think that your children/future generations will have it better or worse than you do to-day?" In other words, compared to studies on environmental awareness and behavior (see currently for Germany, Rubik et al. 2019), a reversed approach was taken: instead of surveying environmentally relevant attitudes and then discussing how these attitudes can translate into behavior (Diekmann and Franzen 2019), particularly committed representatives of local associations, church representatives, teachers, employees in youth and leisure centers, and similar actors were first identified, and only then was it investigated to what extent these change agents displayed openness toward sustainability or climate change mitigation concerns.

The sampling strategy for the twelve interviews (six in each neighborhood) was not based on targeted and chosen combinations of characteristics (such as gender, age, or race of the interviewees), but resulted from the local structures and relevant networks in the research field in the sense of a theoretical sampling (Glaser and Stauss 2017). Recruitment of interviewees was based on research on the Internet and in district newspapers and books, during a site visit, and during participant observation at local events. The interview material was transcribed and then evaluated according to the concept of thematic coding, which serves to verify and further develop theories that are considered promising (Hopf 1993). The most important evaluation categories were constructed inductively from the transcripts in order to obtain a more differentiated picture of which prerequisites and characteristics constitute a potential "change agent for climate protection". On the one hand, it turned out that, in the cases studied, the mental and social prerequisites could be described and that on the other hand, the interviewees had vastly different sustainability-related attitudes.

Social and habitual characteristics of change agents

What conditions enable actors to emerge as local change agents? According to Bourdieu's theory of practice, it is generally possible for actors to help shape or even initiate social change processes through *economic capital*. In the cases studied, the direct use of financial resources by one person (case J, see Table 4.1) played a central role. In this case, a private foundation continued activities that had formerly been carried out by the official state church but could then no longer be financed. However, economic resources can also be of indirect significance. For example, the employee of a youth project in the disadvantaged neighborhood explained the conditions of her local involvement. The economic security granted by her marriage enables her to pursue a profession that is in line with her attitudes but generates a low income.

> Now I don't earn much either. But it's [my income] jointly taxed as a spouse anyway. That's either way. Can't earn much more, practically speaking. (...) Yes of course, that is a certain luxury, to have the time. But if it's enough, if one [within a marriage] earns and one just has to earn a little extra.
>
> (Case A, female, disadvantaged district)

In this case, the economic capital of a spouse could be transferred into time for additional commitment. The interviewee used this time to commit herself beyond the extent that is stipulated in the employment contract. Such indirect recourse to economic capital was found in four of the twelve respondents.

In addition to this reference to an earning spouse, through whom primary economic security is provided, another point is relevant here: the reference to the "luxury of having the time". Having sufficient time at one's disposal was a necessary condition for social engagement in all cases studied. For half of the interviewees, this was made possible by full-time employment; however, some of them also volunteered in the district. Three other actors (cases C, D, and I) were already retired and therefore had the necessary time resources.

Despite the dominant role of economic capital, according to Bourdieu, other types of capital can also serve as a means of power for the actors. *Social capital*, in particular, emerged as an essential resource in the study. Bourdieu (1984) defines social capital as the totality of current and potential resources associated with the possession of a durable network of more or less institutionalized relationships of mutual knowledge or recognition. Several interviewees were part of social networks that enabled them to initiate change. For example, a teacher from the privileged neighborhood (case K) described how she uses project weeks at school to make new contacts in the district that she can use for other activities in the future. The aforementioned employee of a youth project in the disadvantaged district (case A) explains how deeply rooted in the district she has become after ten years of professional experience, how people greet her on the street and children call out to her, which in sum forms the basis of their successful work. For another social worker who used to work in the disadvantaged neighborhood as well, networking with other actors was a central part of his work and he tried to establish relationships with all groups of residents:

> [Our] approach is to work together with everyone. I think it's important that we also get in touch with the citizens from the social institutions, schools and churches [...] That's the linchpin for me.
>
> (Case B, male, disadvantaged district)

Almost all interviewees who were involved in local change processes had distinctive social networks in three different formats: first, a "peer network" of friends and acquaintances anchored in the everyday life, second networking via various institutions in the district, and third, important individuals inside and outside of the neighborhood, the so-called multipliers. This finding is in line with Rogers' (1995) analysis of diffusion processes, which also worked out that social networks, especially close personal relationships, are of central importance for the adoption and thus dissemination of innovations. The fact that economic capital, social capital, and time are mentioned here does not mean that other types of capital, such as symbolic or cultural capital, are not also used as resources for change. However, in the interviews analyzed, we did not specifically ask about the types of capital except for cooperation partners (social capital), and only the three categories of capital discussed here were mentioned.

The endowment with certain resources creates the precondition for an actor to be able to act in an impactful manner. However, this does not mean that they actually act as a change agent. For this reason, we will initially speak only of "change agent potential" in this context. In the case of the people we interviewed, we were able to observe that a habitual disposition must be added to the resources. This disposition is characterized by "wanting to change something" and also actively committing to it. We have tentatively described this disposition as "motivation for change". In the interviews, we repeatedly find passages that express this motivation. A member of a citizens' association for example emphasizes that they implement their projects locally and advocate for changes without external help:

> Even though we have already moved a lot here. But that's really 98 percent of our own work, which we, from the citizens' association, do ourselves.
>
> (Case H, male, disadvantaged district)

Another example is a youth project worker (who was already quoted above), who is described her daily work and stated:

> I do so many home visits here, it's totally effective. (...) And that has also been one of my initiatives. To have as close a connection as possible with the parents.
>
> (Case A, female, disadvantaged district)

Although various social and habitual preconditions can be analytically separated, they often intertwine and are mutually dependent in reality. For example, behind the precondition of "having time", there was, in certain cases, the "taking" of time that was based on a person's motivation for change. Likewise, social capital is generated by engagement and is not a property independent of the actors.

In sum, according to Bourdieu's theory of practice, change agents draw on economic and social capital, in particular, in order to be able to act effectively in their networks. In addition, sufficient time and a motivation to change also plays a central role. The following section will reflect on the extent to which this potential can also be applied to sustainability issues.

Attitudes related to sustainability

In addition to general prerequisites for involvement as change agents, interview questions on the "good life" and intergenerational justice were intended to ascertain the relationship of local actors to sustainability and climate protection goals, thus adding a dimension that has not been examined in Rogers' change agent concept. A large share of interviewees had an explicit sustainability attitude, i.e. a directly affirmative response to the topics of "nature", "environmental and climate protection", and "sustainability", although these topics, as already mentioned, were not directly probed by the interviewers. The interviewees with an explicit sustainability attitude did not necessarily state that they were committed to the protection of the natural environment, but in various contexts, they emphasized its importance or mentioned environmental protection as a component of their commitment.

For example, one interviewee from the privileged district, who was active in party politics and whose work in urban policy committees was primarily concerned with urban planning issues, expressed her preferences in the following way:

> Not cutting back woody plants so often, so that birds can breed there, or not cutting down old trees right away if it's necessary for traffic safety, things like that. Then you can do a lot, with very little actually.
>
> (Case E, female, privileged district)

In both the disadvantaged and the privileged neighborhoods, discussions, and events on the topics of "environmental protection and nature conservation" were organized by churches or institutions close to them (such as the Kolping family), often under the label of "Preserving Creation". Even among change agents who were not directly involved in community work, an orientation toward sustainability in a religious context was evident. For example, the staff member of the youth center (case A), a theologian, tried to convey environmental protection topics pedagogically based on religious ideas (while referencing the Bible and the Koran) during excursions into the forest.

In the case of other interviewees, an attitude was observed that we refer to as an "implicit" sustainability attitude. We attested such an attitude to actors who were open to sustainability and climate protection concerns, or who were already acting in the spirit of sustainability without being directly driven by corresponding motives. For example, an ethos of frugality and the resulting actions (such as reusing plastic bags) do not necessarily aim to protect the environment, but can in fact lead to a more sustainable lifestyle due to the lower consumption of resources. In the case of implicit sustainability attitude, two dominant characteristics could be distinguished:

First, there were interviewees who did not follow a sustainability motive but promoted corresponding activities through their practical actions. In the disadvantaged district, for example, the employee of the youth project regularly organized a bicycling course for migrant women without explicitly pursuing climate protection goals. She reported:

> Some mother from the German course once asked me, can't you teach us how to ride a bike? (...) Then I thought to myself, if there are so

many women, and there are many who can't ride a bike. And it's incredibly useful because it gives women a bit of freedom (...). It gives them self-confidence, the kids realize 'boah, my mom is trying something again'.

(Case A, female, disadvantaged district)

This quote clarifies that the interviewee's commitment was primarily aimed at empowering the migrant women. At the same time, and not intentionally, she also promoted bicycling as a low-emission mode of transportation.

Second, other interviewees showed attitudes which could also advance sustainability if applied consistently. An example of this could be the representative of a church congregation in the privileged district, who advocated the expansion of climate-friendly local public transport, without being motivated by environmental or climate protection goals. Instead, the focus is on revitalizing the district with outside visitors and promoting local businesses. Consequently, regarding the question what would be important for the district in the future, the interviewee answers the following:

The transport connection [of the district] by tram. That would certainly be a positive thing. (...) That perhaps a bit more people stroll here as well.

(Case F, male, privileged district)

The most important topics with an implicit sustainability attitude mentioned in the interviews were: interculturality/integration, religion, promotion of community, and district development/urban planning.

However, there were also a few participants who outright rejected sustainability and environmental protection goals. In both districts, for example, there were citizens' initiatives concerned with local traffic problems. In both cases, the increased car traffic was to be countered with concepts such as moving the tram underground or widening the roads. This is how the chairman of a citizens' initiative in the privileged neighborhood stated:

Even the 70-year-olds are driving more than they used to. That's because they're fitter today. Driving is also getting easier. (...) Long distance [driving] will perhaps not grow so much, but short distances will. And if the city of Cologne doesn't take this into account in its transport planning, then it will affect its economic performance or its competitiveness, and then people will move elsewhere.

(Case D, male, privileged district)

The same person also generally rejected climate and environmental protection as "emotionally overblown" issues and denied anthropogenic causation of climate change. In addition, a few other interviewees (cases C, G, and J) did not reveal any particular attitude, neither positive nor negative, toward sustainability issues. These observations show that different points of entry for sustainability issues and climate protection exist. In addition, we identified a number of implicit links to sustainability in the interviews via other locally important topics. In this sense, unusual suspects who can act as change agents for climate protection and influence the local culture in neighborhoods – even without an explicit pro-environmental attitude – were identified.

Overall comparison

In the data analysis, the presence of social and/or economic capital, as well as the availability of time, emerged as central prerequisites for the commitment and social effectiveness of local change agents. In addition, there was a pronounced habitual willingness to initiate or participate in change processes. Almost half of the respondents were explicitly open to climate protection or sustainability issues. Other change agents revealed an implicit affinity for more environmentally friendly social practices, without directly pursuing climate protection or sustainability goals. Based on the limited data that was collected, we cannot conclude that an implicit sustainability attitude also results in openness toward climate and environmental protection measures, or even actual efforts regarding environmental protection.

Nevertheless, there are several corresponding links and potentials. Table 4.1 shows an overview of all cases, indicating the respective social and habitual preconditions, as well as sustainability-relevant attitudes.

Three groups can be distinguished: first, there is a majority of local change agents who are explicitly or implicitly enthusiastic about sustainability issues or already act accordingly (group 1, case A, B, E, F, H, K, L). Second, there are some actors who have only a limited motivation to change and/or who do not reveal, either implicitly or explicitly, a sustainability-relevant attitude (group 2, case C, G, J). Lastly, there is a small group of actors who are, more or less explicitly, hostile to climate protection and sustainability concerns (group 3, case D, I).

Discussion

Our findings show the importance of actors, who are well integrated in socio-spatial networks, in creating change or influencing local discourses on sustainability and climate protection. Rogers (1995) already pointed out that while mass media are suitable for making an innovation known, social networks and especially close personal relationships are crucial for the individual adoption of innovations. Social capital was also the most crucial resource of almost all the change agents studied here. In disadvantaged neighborhoods, the central actors are strongly networked through professional structures in the field of social work, which opens up particular opportunities for change.

Table 4.1 Characteristics and sustainability-related attitudes of actors
Group 1: Change agents with an affinity toward sustainability. Group 2: Actors with limited motivation for change and/or without clear reference to sustainability issues. Group 3: Change agents with rather negative attitudes toward climate protection and sustainability goals

Neighborhood	Case	Group	Engagement		Social and habitual conditions					Motivation for change		Attitudes toward sustainability			
					Economic capital		Social capital						Implicit		
			Volunteer	Full-time staff	Direct	Indirect	Peers	Institutional	Multiplicator(s)	Motivation for change	Time	Explicit	Attitude	Practice	Aversion
Disadvantaged neighborhood	A	1	x	x		x	x	x		x	x	x		x	
	B	1		x				x	x	x	x	x			
	H	1	x				x	x	x	x	x		x		
	L	1	x	x			x	x	x	x	x	x		x	
	C	2	x				x				x			x	
	G	2		x			x	x		x	x				
Privileged neighborhood	E	1	x			x	x	x		x	x	x	x		
	F	1		x			x			x	x		x		
	K	1	x	x			x	x	x	x	x	x		x	
	J	2	x		x		x		x	x	x				
	D	3	x		x	x		x	x	x	x				x
	I	3	x			x	x	x	x	x	x				x

Source: Own representation.

Furthermore, the results of the study indicate which role local change agents could play in the formation of different climate cultures. In particular, it became apparent that beyond known pro-environmental actors, there are local change agents in communities who are open to the goals of climate-friendly and sustainable urban development and spread these ideas and practices in their networks. Different kinds of knowledge constructions concerning possibilities and restrictions in dealing with climate change are shared by these change agents, who have internalized specific knowledge in their past about what restricts "the scope of future interpretation and action" (Heimann and Mallick 2016, 59).

The local anchoring of the actors in their neighborhoods makes it plausible "that orders of knowledge can take shape not only historically with the passage of time, but also socio-spatially at a specific moment in time" (Heimann 2019, 17). Due to different role requirements in functionally differentiated societies, successful socialization does not mean that attitudes and actions always coincide, but rather that people are able to deal flexibly with different requirements. In this context, it is not the territorial area of the urban district that was decisive, but rather the networks situated there (applying a relational concept of space: Heimann 2019; Löw 2016). This can be seen, for example, in the fact that some of the change agents studied here, such as committed church representatives, social workers, or sports club leaders, were not residents of the district, but worked there, maintained intimate relationships with residents, and spent a large part of the working hours in the neighborhood. Their knowledge of and attitudes on environmental issues were formative in shaping local engagement with sustainability initiatives. It stimulated willingness to participate by others involved in the district and by residents. Emerging climate cultures can not necessarily be characterized by a general change in environmental awareness but involves particular practices, which are implemented in everyday life.

The presented study had a strongly exploratory character and comparatively low number of cases. More extensive in-depth analysis in different regions and neighborhoods promises to provide further findings on emerging local climate cultures and conditions of their genesis.

Note

1   This chapter is a revised and expanded version, translated into English, of a paper published in German in 2014 by the same authors (Sommer and Schad 2014).

References

Beckhard, Richard. 1969. *Organization Development: Strategies and Models.* Reading, MA: Addison-Wesley.

Bliesner, Anna, Christa Liedtke, and Holger Rohn. 2013. "Change Agents für Nachhaltigkeit: Was müssen sie können?" [Change Agents for Sustainability: What Do They Need to Be Able to Do?]. *Zeitschrift Führung + Organisation* 82 (1): 49–53.

Böcker, Maike, Henning Brüggemann, Michaela Christ, Alexandra Knak, Jonas Lage, and Bernd Sommer. 2021. *Wie wird weniger genug? Suffizienz als Strategie für eine nachhaltige Stadtentwicklung* [How Does Less Become Enough? Sufficiency as a Strategy for Sustainable Urban Development]. München (Germany): oekom.

Bourdieu, Pierre. 1977. *Outline of a Theory of Practice.* Cambridge (England): Cambridge University Press.

Bourdieu, Pierre. 1984. *Distinction: A Social Critique of the Judgement of Taste.* Cambridge, MA: Harvard University Press.

Clar, Christoph, and Reinhard Steurer. 2019a. "Climate Change Adaptation at Different Levels of Government: Characteristics and Conditions of Policy Change". *Natural Resources Forum* 43 (2): 121–131.

Clar, Christoph, and Reinhard Steurer. 2019b. "Climate Change Adaptation Strategies at Different Levels of Government". In *Research Handbook on Climate Change Adaptation Policy,* edited by E. C. H. Keskitalo and B. L. Preston, 310–326. Cheltenham (England): Edward Elgar Publishing.

Diekmann, Andreas, and Axel Franzen. 2019. "Environmental Concern: A Global Perspective". In *Einstellungen und Verhalten in der empirischen Sozialforschung: Analytische Konzepte, Anwendungen und Analyseverfahren* [Attitudes and Behavior in Empirical Social Research: Analytical Concepts, Applications, and Analytical Procedures], edited by Jochen Mayerl, Thomas Krause, Andreas Wahl, and Marius Wuketich, 253–272. Wiesbaden (Germany): Springer Fachmedien Wiesbaden.

Elias, Norbert. 2001. *The Society of Individuals.* New York: Continuum.

Fichter, Klaus, and Jens Clausen. 2013. *Erfolg und Scheitern 'grüner' Innovationen: Warum einige Nachhaltigkeitsinnovationen am Markt erfolgreich sind und andere nicht* [Success and Failure of 'Green' Innovations: Why Some Sustainability Innovations Succeed in the Marketplace and Others Do Not]. Marburg (Germany): Metropolis-Verlag.

Giddens, Anthony. 1986. *The Constitution of Society: Outline of the Theory of Structuration.* Cambridge (United Kingdom): Polity Press.

Heimann, Thorsten. 2019. *Culture, Space and Climate Change: Vulnerability and Resilience in European Coastal Areas.* London (United Kingdom): Routledge.

Heimann, Thorsten, and Bishawjit Mallick. 2016. "Understanding Climate Adaptation Cultures in Global Context: Proposal for an Explanatory Framework". *Climate* 4 (4): 59.

Hoffman, Andrew John. 2010. "Climate Change as a Cultural and Behavioral Issue: Addressing Barriers and Implementing Solutions". *Organizational Dynamics* 39 (4): 295–305.

Hopf, Christel. 1993. "Soziologie und qualitative Sozialforschung" [Sociology and Qualitative Social Research]. In *Qualitative Sozialforschung* [Qualitative Social Research], edited by Christel Hopf, and E. Weingarten, 11–37. Stuttgart (Germany): Klett-Cotta.

Hopkins, Rob. 2013. *The Power of Just Doing Stuff: How Local Action Can Change the World.* Cambridge (United Kingdom): Green Books.

Kristof, Kora. 2010. *Wege zum Wandel: Wie wir gesellschaftliche Veränderungen erfolgreich gestalten können* [Paths to Change: How We Can Successfully Shape Social Change]. München (Germany): oekom.

Kristof, Kora. 2020. *Wie Transformation gelingt: Erfolgsfaktoren für den gesellschaftlichen Wandel* [How Transformation Succeeds: Success Factors for Social Change]. München (Germany): oekom.

Kristof, Kora. 2021. „Erfolgsfaktoren für die gesellschaftliche Transformation: Erkenntnisse der Transformationsforschung für erfolgreichen Wandel nutzen" [Success Factors for Societal Transformation: Using Insights from Transformation Research for Successful Change]. *GAIA – Ecological Perspectives for Science and Society* 30 (1): 7–11.

Leggewie, Claus, and Franz Mauelshagen. 2018. "Tracing and Replacing Europe's Carbon Culture". In *Climate Change and Cultural Transition in Europe*, edited by Claus Leggewie and Franz Mauelshagen, 1–22. Leiden (Netherlands): Brill.

Leggewie, Claus, and Harald Welzer. 2010. *Das Ende der Welt, wie wir sie kannten: Klima, Zukunft und die Chancen der Demokratie* [The End of the World as We Knew It: Climate, Future and the Chances of Democracy]. Frankfurt am Main (Germany): Fischer.

Lloyd, Christopher D., Ian G. Shuttleworth, and David W. S. Wong. 2014. "Introduction". In *Social-Spatial Segregation: Concepts, Processes and Outcomes*, edited by Christopher D. Lloyd, Ian G. Shuttleworth, and David W. S. Wong, 1–10. Bristol (United Kingdom): Bristol University Press.

Löw, Martina. 2016. *The Sociology of Space: Materiality, Social Structures, and Action*. New York: Palgrave Macmillan.

Mautz, Rüdiger, Andreas Byzio, and Wolf Rosenbaum. 2008. *Auf dem Weg zur Energiewende: Die Entwicklung der Stromproduktion aus erneuerbaren Energien in Deutschland* [On the Way to the Energy Transition: The Development of Electricity Production from Renewable Energies in Germany]. Göttingen (Germany): Universitätsverlag Göttingen.

Rogers, Everett M. 1995. *Diffusion of Innovations*. New York: The Free Press.

Rubik, Frieder, Ria Müller, Richard Harnisch, Brigitte Holzhauer, Michael Schipperges, and Sonja Geiger. 2019. *Umweltbewusstsein in Deutschland 2018: Ergebnisse einer repräsentativen Bevölkerungsumfrage* [Environmental Awareness in Germany 2018: Results of a Representative Population Survey], broshure edited by Bundesministerium für Umwelt, Naturschutz und nukleare Sicherheit (BMU) and Umweltbundesamt (UBA). Berlin (Germany). https://www.ioew.de/publikation/umweltbewusstsein_in_deutschland_2018

Sommer, Bernd, and Miriam Schad. 2014. "Change Agents für den städtischen Klimaschutz. Empirische Befunde und praxistheoretische Einsichten" [Change Agents for Urban Climate Change Mitigation. Empirical Findings and Insights from Practice Theory]. *GAIA – Ecological Perspectives on Science and Society* 23 (1): 48–54.

Welzer, Harald, Hans-Georg Soeffner, and Dana Giesecke, eds. 2010. *KlimaKulturen: Soziale Wirklichkeiten im Klimawandel* [Climate Cultures: Social Realities in Climate Change]. Frankfurt am Main (Germany): Campus.

WGBU – German Advisory Council on Global Change. 2011. *World in Transition. A Social Contract for Sustainability. German Advisory Council on Global Change*. Berlin (Germany). https://www.wbgu.de/fileadmin/user_upload/wbgu/publikationen/hauptgutachten/hg2011/pdf/wbgu_jg2011_en.pdf

Zell-Ziegler, Carina, Johannes Thema, Benjamin Best, Frauke Wiese, Jonas Lage, Annika Schmidt, Edouard Toulouse, and Sigrid Stagl. 2021. "Enough? The Role of Sufficiency in European Energy and Climate Plans". *Energy Policy* 157: 112483.

5 Fractured Climate Cultures in Depopulated Southern Spanish Communities

Pilar Morales-Giner

Introduction

Recent social movements such as "the Revolt of the Empty Spain" and "Farmers to the Edge" underscore the struggles of communities in Spanish rural areas. These social movements formed in response to a series of ongoing environmental, economic, and demographic changes that date back to the middle of the 20th century.

Researchers have identified changing climatic trends during the 20th and 21st centuries in Spain. Experts note a clear increase in temperatures, higher aridity, some decrease in overall rainfall, and an increase in rain intensity (Paniagua et al. 2019; de Castro et al. 2004). Several other studies have confirmed changing climatic trends in southern Spain during the last century, i.e. decrease in precipitation and increase in the severity of droughts (Peña-Gallardo et al. 2016; Sinoga-Ruiz et al. 2011). Moreover, some of the areas in the province of Granada, such as Sierra Nevada and Sierra de Baza, report higher decrease in precipitation compared to other regions in Spain (Peña-Gallardo et al. 2016). These trends are projected to continue and even intensify in the near future as some regions could see a 10% decrease in mean winter precipitation and an increase of 1°C–2°C in summer temperatures for 2011–2050 (Coll et al. 2015). Trends such as increasing aridity and rain variability lower the climatic suitability for agricultural activities (Paniagua et al. 2019; Sinoga-Ruiz et al. 2011). In fact, recent research has shown how climatic variation is already having an effect on flowering periods of olives in southern Spain, i.e. lengthening the pollen season (García-Mozo et al. 2014).

Alongside climatic shifts, rural areas have been exposed to a series of socio-economic changes such as depopulation, deagrarianization, and deterritorialization. Many rural areas in Spain have experienced a steady demographic decline. This decline has recently been highlighted in the national political and public spheres through activism and political representation in the national institutions. However, Infante-Amate et al. (2016, 74) note how the depopulation of rural Spain started to take place at a systemic level in the 1960s as "a result of mechanization of agricultural work and the rise of cities". In the 1950s, 39% of the Spanish population lived in villages of

DOI: 10.4324/9781003307006-9

fewer than 2,000 inhabitants. By 2017, this percentage had decreased to 18% (Pinilla and Saez 2017). Furthermore, globalization and industrialization processes have led to a series of severe consequences such as loss of inland farms, increase of coastal greenhouse farms, reliance on migrant labor, income decline, high reliance on domestic and European Union subsidies, and ongoing decline of the agrarian aging population struggling to find generational replacement. Some of these trends, such as income decline, are particularly notable for small-scale farmers, which in turn often leads to either farm abandonment or intensification of agriculture to increase profitability (González de Molina et al. 2019; Cabello and Castro 2012).

Despite this process of intensification of agriculture in the area, there are two specifications of agrarian population in southern Spain worth mentioning. First, the agricultural sector is stronger in Andalusia compared to other regions in Spain. For example, in 2016, the agricultural sector GDP in Andalusia was 7.1% of the total GDP in the region, more than double the national average (3.1%) (INE 2019). Therefore, small-scale agriculture is still widespread and linked to part-time holders who are often elderly and male (Ortiz-Miranda et al. 2013). Cabello and Castro (2012) claim that small-scale traditional agriculture in Andalusia, developed through centuries, has allowed the conservation of native crops. Madejón et al. (2011) present evidence of how traditional soil management practices in southern Spain have enabled sustainable remediation of severely polluted soils. Furthermore, Gómez-Baggethun et al. (2012) show how traditional knowledge in southwestern Spain has helped maintain community resilience to environmental hazards, most notably droughts. These authors along with Cabello and Castro (2012) express concern over the current loss of traditional knowledge in rural areas. Second, the last decades have shown an acceleration of reforestation as well as a decrease in some indicators of intensification, e.g. fertilizer consumption (González de Molina et al. 2019; Ortiz-Miranda et al. 2013).

Still, the literature clearly points to how the intensification of agricultural systems in Spain, including the South, has come at the expense of cultural, social, and ecosystem services (Cabello and Castro 2012). Thus, while agrarian communities in depopulated southern rural areas maintain some degree of sustainability, they also face unique hardships due to environmental, economic, and demographic processes.

In the following sections, I cover the main theoretical underpinnings of this analysis: sense of place, political ecology, and climate cultures. Then, after outlining my methodological strategy, I explain my study's main results and their implications. Ultimately, I argue that in depopulated southern Spanish communities, there are conflicting identities and social processes that denote a fractured climate culture.

Climate cultures in shadow landscapes

Social science research, especially in human geography, has long incorporated sense of place, understood as notions of spaces that construct cultural and social identities. Indeed, place is a relevant category for social science, as Gieryn argues that *places* are *spaces* "filled up by people, practices, objects, and representations" (Gieryn 2000, 465). However, until recently, "place" as a dimension of human responses to climate change has been neglected (Masterson et al. 2017; Devine-Wright 2013). The notion of climate cultures incorporates spatial references into the discussion on collectively shared knowledge about both climate vulnerability, which includes perceptions about climate change and resilience (Heimann 2018). Adger (2016) shows that issues such as the experience of environmental hazards strengthen connections between perceived weather patterns and climate change, which are interpreted through values relating to familiar places. Tschaker et al. (2019) conduct a systematic literature review on place-specific experiences of climate change that empirically validates the harm in relation to intangible aspects of loss, such as homesickness, loss of identity, etc. Devine-Wright (2013) maintains that threats to sense of place may lead to a greater willingness to act and adapt. Conversely, Marshal et al. (2012) point to how notions of place and occupational identity can act as barriers to adaption strategies. Place then serves as a catalyst of climate change perception with potential to motivate and limit action.

Experience of environmental hazards, tangible and intangible loss, and perceptions of place in relation to climate are different dimensions of what Heimann (2018) calls *relational cultural spaces* that carry knowledge about responses to climate change. Indeed, both local communities and climate are linked to their spaces. Thus, adopting a relational climate culture lens requires investigating "a 'mingling' of place, personal history, daily life, culture, and values" (Brace and Geoghegan 2011, 286). In other words, a relational climate culture involves understanding how place shapes communities' identities, otherwise known as *place identity*.

Place is understood as a social construct in which social actors create place identity via a series of socio-economic and cultural processes in their local spaces. Through these processes, local communities territorialize their space and create a shared and distinct concept of place (Entrena Durán 2010). At the same time, these spaces are embedded in contextual power dynamics such as state and regional policies, market demands, and global capitalism. In the context of depopulated areas in Southern Europe, Bryant et al. (2015, 233) coined the term "shadow landscapes", "which brings together processes of marginality, scale, socio-nature, and cultures of depopulation to explain human–environmental dynamics in those areas marked by the relative absence of people". Shadow landscapes are capable of illuminating relational cultural spaces that carry knowledge about responses to climate change.

In rural areas in southern Spain, depopulation, deagrarianization, and deterritorialization are key generators of shadow landscapes. Depopulation refers to steady demographic decline that can lead to several problems, including difficulties in maintaining public services and decreasing tax revenues (Elshof and Bailey 2015). "Deagrarianization" – that is, a breakdown between agriculture and rural territories (del Pino and Camarero 2017) – consists of diminishing crop yields and livestock multifunctionality, which leads to the intensified exploitation of croplands, labor flexibilization, increase in the use of water resources, and outsourcing, including labor, energy sources, and biomass (González de Molina et al. 2019; Ortiz-Miranda et al. 2013). Small towns are also experiencing a process of deterritorialization, according to which social relations as well as identities increasingly depend on factors determined elsewhere, e.g. in urban areas, global markets, and regional governments (Entrena Durán 1999). At the same time, these areas still maintain cultural distinctiveness illustrated, among other things, through agroecological traditions. As Bryant et al. (2015, 236) put it, "shadow landscapes" reflect "culturally distinctive if often opaque and shifting geographies that are irreducibly human creations marked by deep ambiguity and emotional attachment".

As noted above, demographic and environmental changes in Andalusian rural areas occurred alongside systemic transitions, e.g. market demands and changes in the production system. The climate cultures literature maintains that actors develop shared knowledge constructions that, in turn, form relational cultural spaces. I argue that in depopulated towns in rural Andalusia, shared knowledge constructions are shaped around shadow landscapes. An examination of shadow landscapes will reveal the extent to which responses to climate change – in particular, perception and adaptation – are shaped by the relationships between inhabitants and spaces while at the same time attending to how the social construction of places reproduces and sometimes challenges existing power dynamics.

Study area and methods

The fieldwork consisted of 24 semi-structured interviews conducted during the summer of 2019. The aim of the study was to reach out to populations living in small rural towns that are depopulating, at least a loss of 2% since 2000 (see Table 5.1). I visited a total of ten municipalities in five *comarcas* (subregional rural areas). The population in these towns tends to be older than the regional average (see Table 5.1). In addition, their agricultural sector tends to be significant. For example, towns in Table 5.1 have higher rates of n. of contracts between employee and employer in the agricultural sector compared to the regional rate. Note that these contracts overwhelmingly are temporary. The local unemployment rate also is higher than regional and national averages.

Participants were recruited through personal contacts and snowball sampling. The only requirement to participate in the study was that the candidate knew the town well. Preference was given to individuals who were more likely to have a close relationship to the socio-natural spaces in the town, i.e. those who work or have worked in the agricultural sector as well as individuals who are or have been involved with the town's local

Table 5.1 Social characteristics of towns compared to region (Andalucía)

Municipalities	Comarcal/Leader area	Total population 2018	Population change 2000–2018 (%)	Age average 2018	Unemployment rate 2018 (%)	Agricultural contracts 2018 (%)	Temporal agricultural contracts 2018 (%)
Algarinejo	Poniente Granadino	2,591	–51.0	49.4	23.17	82	100
Alpujarra de la Sierra	Alpujarra- Sierra Nevada de Granada	978	–18.0	49.4	20.59	48	100
Arenas del Rey	Poniente Granadino	1,828	–8.0	47.5	23.06	77	100
Gor	Guadix	736	–36.0	55.8	21.65	30	100
Guajares, Los	Valle de Lecrin- Temple & Costa	1,028	–20.0	51.1	23.18	34	100
Iznalloz	Montes de Granada	6,775	–3.0	38.6	27.01	45	100
Lecrin	Valle de Lecrin- Temple & Costa	2,103	–8.0	47.5	19.14	55	100
Pedro Martinez	Guadix	1,134	–17.0	47.8	28.86	68	100
Válor	Alpujarra- Sierra Nevada de Granada	672	–27.0	51.4	24.7	25	100
Villanueva de las Torres	Guadix	585	–35.0	47.6	32.17	43	100
Andalucia		8,384,408	14.0	41.6	23.39	30	98

Sources: Junta de Andalucía 2020, INE 2020

administration. Such individuals are more likely to be familiar with the demographic, socio-economic, and climatic changes in the towns. Interviews
were conducted in Spanish and relevant quotes were translated into English.
In order to protect participants' privacy and following IRB (Institutional Review Board) guidelines, the names assigned to each participant do not correspond with their real names. Among the respondents included were farmers,
retirees, political leaders, and church leaders (see Table 5.2). Before each interview, I informed participants of my background as a woman that grew up
in Southern Spain and who is now pursuing a PhD program in the United
States. I conducted the coding analysis applying audio-coding, a method that
allows the researcher to stay sensorially closer to the original data (Wainwright and Russel 2010), and following broad theoretically driven categories,
e.g. sense of place, climate change perception, and adaptation.

Table 5.2 Semi-structured interviews (participants)

Participant's name	Municipality (involved with or lives in)	Gender	Age	Occupation
Javier	Algarinejo	M	52	Agricultural Sector Worker
Eduardo	Algarinejo	M	43	Farmer
Mateo	Alpujarra de la Sierra	M	68	Retired/Farmer
Francisco	Arenas del Rey	M	55	Local Administration/ Farmer (part-time)
Sara	Arenas del Rey	F	49	Local administration
Sergio	Arenas del Rey	M	34	Service Sector
Carlos	Gor	M	87	Retired/Farmer
Julián	Gor	M	73	Retired/Service Sector
Mario	Gor	M	74	Retired/Service Sector
Hugo	Gor	M	79	Retired/Farmer
Pedro	Iznalloz	M	49	Farmer/Business Owner
Daniel	Iznalloz	M	59	Service Sector/Business Owner
Laura	Iznalloz	F	35	Service Sector
Ángel	Lecrín	M	53	Service Sector Worker
David	Lecrín	M	42	Farmer/Business owner
Sonia	Los Guajares	F	66	Retired
José	Pedro Martinez	M	65	Retired/Local Administration
Joaquín	Pedro Martinez	M	74	Local Administration
Oscar	Pedro Martinez	M	73	Retired/Local Administration
Martín	Pedro Martinez	M	78	Retired/Farmer (part-time)
Nestor	Pedro Martinez	M	84	Retired/Service Sector
Pablo	Pedro Martinez	M	51	Local Administration/ Farmer (part-time)
Luis	Válor	M	62	Farmer
Rodrigo	Villanueva de la Torre	M	49	Farmer (part-time)

Source: Own representation.

While conducting the analysis of the interviews, the categories "loss" and "abandonment" emerged inductively. These categories have noted to be relevant both in the context of climate change (Tschakert et al. 2019) as well as in the in the context of depopulated rural towns (Bryant et al. 2015). Thus, to achieve a more generalized picture of loss and abandonment in rural depopulated areas, I conducted a content analysis of LDS (local development strategies) for the period of 2014–2020 supported by LEADER, a French acronym that in English refers to links between rural economy and development actors. LEADER is an EU initiative to alleviate inequalities in rural areas of member states (European Commission 2014; Junta de Andalucía ND). Through these programs, rural development groups prepare LDS through a participatory methodology in which they engage local communities and stakeholders (European Commission 2014).

I analyzed the LDS documents that pertain to the five *comarcas* of the municipalities I visited. This process included two steps. First, 133 quotes that contained the words "loss" or "abandonment" were identified and coded. These quotes referred to observed trends and threats in relation to loss or abandonment in the rural region. Second, the codes were grouped into five categories to capture the perceived adverse changes taking place in these rural areas. These categories of loss provide a picture of the spaces where shared knowledge constructions about climate change take place.

Drawing from the analysis of the interviews and the LDS documents, the next section shows how inhabitants of rural depopulated areas exhibit attachment to spaces that are experiencing several degrees of change, including loss, deagrarianization, deterritorialization, and some recovery efforts. The two sections that follow rely on the theoretical frameworks of sense of place and political ecology to describe responses to climate change in the study areas. Understanding these responses is relevant to identifying the nuances of climate culture in depopulated rural areas.

Place identity in depopulated and changing Southern Spanish communities

In this section, I cover two major contextual themes that emerged in the analysis and help understand place identity in the areas I visited. First, I examine the high attachment of participants to their towns. Then, I describe how these areas experience different degrees of change, loss being the most drastic form. This section helps situate the reader in the context of the depopulated towns studied before turning to responses to climate change.

Participants revealed a high attachment to their towns. Indeed, towns were often described as *calm* and *peaceful*. In these, inhabitants develop a healthy and comfortable lifestyle while also maintaining a sense of community through abundant social ties. Participants often articulated these attributes in opposition to urban life, defined as a stressful environment. Some of the respondents attributed high identification with their town. At times, participants referred to towns as their roots or used possessive adjectives when referring to the town and what it constitutes, e.g. my land or my trees. In fact, participants talked about how they stayed in their towns because they identified with the land. Also, some of those who leave come back when they retire because of homesickness. This identification with the town also applied to the occupations typical of this region, namely farming. Participants specifically referred to farming as a satisfactory activity that suits them and thus placed high value on the town's ecological resources: climate, air, and landscape quality.

This narrative of rural towns as idealized or unchanging socio-natural spaces is often further emphasized by external and local actors. Urban areas inhabitants, who practice short-term rural tourism or who own a secondary residence in rural areas, tend to see these rural spaces as "exotic throwbacks to the long-lost past" (Bryant et al. 2015, 241). In addition, features of rural depopulated towns, such as ageing and small populations and strong primary sector, can stimulate a picture of towns as unchanging places. For example, one of the respondents referred to his town as a place in which time "goes by slowly with the same people and same stories". However, rural towns continue to be exposed to accelerating shifts such as depopulation, deagrarianization, and deterritorialization. These shifts are often "linked to multi-scalar politics beyond their control" (Bryant et al. 2015, 242). In accordance with these environmental and demographic changes, inhabitants of depopulated areas face marginality that is discernible through the processes of loss and abandonment. Indeed, content analysis of the LDS documents and interviews showed that "loss" and "abandonment" were quite present (133 times).

Five consistent interrelated themes emerged that will be discussed in order of higher to lower frequency of mention.

1   *Agricultural & Natural Ecosystems.* Many of the resources lost or abandoned are essential to the functioning of agricultural and natural ecosystems, including agricultural products and their value, land for cattle, and traditional agricultural management. The LDS reports also mentioned losses in relation to the natural ecosystem: soil (due to erosion), water, biodiversity, and reforestation efforts that suffer from abandonment. Participants often mentioned the abandonment of traditional strategies in favor of the modernization of irrigation systems and the intensification of land treatments. Farmers also recounted that they had eliminated the multifunctional use of livestock in favor of motorized machinery for tillage. In addition, the interviews revealed loss of water in canals, fountains, and soils. *Luis* claimed that there has been a loss of "freshness" because it rains less and at the same time water runs through tubes in irrigation systems, drying the soil and the vegetation therein.

2 *Culture, Traditions, & Heritage.* While perceptions of loss mostly involve traditional agricultural practices, they also include cultural and traditional activities linked to the lack of generational replacement. The interviewees specifically referred to loss of culture linked to water. For example, mountain areas are losing their "acequias de careo", which allow for the redistribution of water through the year without building dams (Espín Piñar et al. 2010). *Luis* and *Eduardo* indicated how these traditional water management canal systems lack both people who manage them and sufficient water supply. *David* explains that now there is "less *entertainment* of water" in the soil of the mountains because less people irrigate and therefore the water flows to the sea and is lost. He explains how water was previously *entertained* during spring generating a larger supply in aquifers and springs. *David* explained that there used to be more respect for water as it was used to irrigate the food they ate. Now the water is used to grow crops that are later sold and processed.

3 *Services, Employment, and Fair Labor Practices.* Basic public services as well as businesses, including small retailers and co-ops, are disappearing or at risk of disappearing in these areas. Likewise, the interview participants lamented low accessibility to basic services, such as education, healthcare, infrastructure. Some mentioned that local job opportunities are not considered attractive by inhabitants as the jobs often go to immigrants. However, several participants pointed to the loss of employment and job insecurity motivated by the mechanization of agriculture, most notably, the dependence on risky harvests with low market value. These *risks* are particularly prominent for dryland farms that depend on rainfall decisively. But hail and frosts are also cited as risks. Participants also suggested that their work, despite its difficulty, is not valued. *Javier* complains that small-scale farmers cannot "put a price" on their labor as the costs of production often outweigh the benefits. In fact, occasionally landfills are seeded, but the crop is not collected due to a lack of profitability. Therefore, in the view of several participants, the lack of steady income deprives the youth from affording basic needs to start a family.

4 *Population.* LDS documents often mentioned demographic loss in relation to the exodus of youths and women. Demographic loss was further attributed to brain drain, decrease in public investment, social drive, tradition changes, and population flight. Population loss was also addressed as a prominent worry during my fieldwork. In one instance, a restaurant owner acknowledged that business was not going well due to loss of citizens. To exemplify the demographic decline, she noted that five people had died the previous week.

5 *Place.* The content analysis revealed the materiality of what is being lost or abandoned: land, crops, facilities, soil, and built environment such as infrastructure, hiking paths, etc. The abandonment of places

also came up in interviews. For example, *Laura* explained how she noticed a lot of empty houses, "I don't like going out and seeing the abandonment", she says. Others referred to countryside abandonment. Pablo says, "your soul drops to the floor, the trees are lost (...) all is completely lost, it is like a desert". In this way, abandoned spaces lack the physical presence of people as well as interactions between people and space. Ultimately, the loss of physical as well as social places epitomizes shadow landscapes that are facing accelerating economic and environmental shifts.

Admittedly, not everything has been lost in these places. Some traditional agricultural practices are being rescued alongside the adoption of agroecology in some rural areas (González de Molina et al. 2019). In addition, one of the participants mentioned a project to build a solar farm in his town. This project was framed as 'dream' that could bring hundreds of jobs to the town. Different administrative units are also making some effort to maintain and develop rural areas, for example, through rural development groups, conservation initiatives, and external public funds mostly provided by the Common Agrarian Policy of the European Union and the Spanish government. The Agrarian Unemployment Subsidy also provides several months of unemployment benefits to farm workers that declare a set amount of working days (Suárez-Navaz 2004). While some acknowledged the crucial role of these funds for the survival of agriculture, others expressed disapproval toward such subsidies as in their view they promote indolence.

In addition, some of these conservation and development efforts are at times perceived as 'deterritorialized' as they benefit the population in urban areas. *Mateo* reports his experiences from a meeting about projected celebratory events for a natural park. The events were all proposed to take place in bigger cities, excluding the municipalities in which this natural park is situated. *Mateo* raised concerns about this issue during the meeting. Although some of the attendees agreed with him, he expressed skepticism over plans being actually changed: "it is always said, and never done".

Despite these efforts to maintain rural areas, concerning shifting structural conditions such as intensification, industrialization, and globalization have provoked a series of farmers' mobilizations during 2019 and 2020 under the motto "farmers on the edge". Among other things, these movements denounce unfair retail prices, subsidy cuts, precarious labor conditions, trade barriers resulting from Brexit and new US tariffs, lack of wildlife control, and climate change impacts (UPA 2020). Indeed, the working conditions make the job hard to bear. *David* describes the physical and social difficulties of farm labor, admitting that sometimes he only sees three people a day and that this makes him feel overwhelmed when he visits crowded spaces. *Mateo* expresses this sentiment with indignation: the fact "that I like the countryside doesn't mean I need to be a slave, and get used to not living well, or [that I] condemn my family". In addition, *David* explained, not everyone can mechanize in the same way, as farmlands in mountainous terrain pose physical obstructions to mechanization, and respective farmers 'can't compete with machines'. *Eduardo* goes on to explain that

> we are competing in a world in which they are doing very intensive olive tree agriculture (...) they are using a harvester (...). Here we have olive trees that are 500 or 600 years old, but [the olive trees in intensive farms] that are four or five years (...) don't need work force, it is all done by a machine. Our towns provide a lot of employment, it is concerning.

The study areas present some degree of fracture in their socio-natural landscape. Indeed, the analysis of the interviews and LDS documents reveals recurring sifts in shadow landscapes: deagrarianization, depopulation, and deterritorialization. These processes are discernible through several layers of loss and abandonment. Despite these dynamics of marginality, some agricultural and traditional practices are being protected and rescued.

Climate change perception in depopulated communities

Perception of impacts and vulnerabilities around climate change

Social constructions about place are key components of *knowledge systems* around climatic changes. Indeed, respondents used place as a point of reference to communicate perceived environmental changes. To illustrate climate change, *Joaquín* referred to decreasing amounts of birds in the town. Two participants indicated that due to their place-bound occupation, farming, they are aware of the changes in the weather. *David* explained that when he wakes up, he looks at the sky.

Typical points of reference were centered around water, snow, and ice. Participants mentioned dry streams and canals; fountains and cracked land near river basins; and experiences with snow and ice that are now gone. Such loss experiences include snow in the town's plaza, shoveling snow, being buried in snow, ice-skating without skates, and frozen clothes on the clothesline. When I asked *Pablo* about the impacts of climate change, he referred to how the lack of rain deeply affected crops in dryland areas. He explains that he lost 1,400 almond trees and not even one persisted because of the lack of water. Another farmer referred to rain as the food of soil and the life of everything. He describes that "if it doesn't rain [the soil] doesn't have its food". Pablo also reported how this lack of water is a stronger problem than it used to be as it continues through consecutive seasons. "Before (...) we could recover", he says. Some also expressed how the lack of water is now more continuous and, therefore, intense, for example Mario says: "when I was little, I remember there were years of dryness, but not like now, not that many consecutive years. We used to live from the countryside, and now we cannot do that anymore".

This relational understanding of climate change was consistent with the climatic changes described in the literature, participants perceived a general increase in temperatures, and decreased but more intense rainfall. Some respondents also reported shifts in seasons, e.g. longer and earlier summers. Older participants tended to recognize longer-term climatic changes. For example, *Joaquín* said:

> climate change from 40 years to now, you can see it, one day and another day, it rains less, it snows less, there are less birds, way less. I remember when I was young, there were many different varieties of birds, sparrows there are quite less, there are [almost no swallows] coming.

The effects of more intense storms affecting specific places were also mentioned. *Rodrigo* reports how recently 'catastrophic' rains cause major destruction. Another farmer recounted how the land is being eroded by more frequent torrential rains, remembering olive trees uprooted near the river.

When observing differences in perception about climatic changes, the dimensionality of relational cultural spaces is exposed. For example, *Julián*, who was in a town that lacked access to abundant water for decades, explains that he has only noticed a slight decrease in rainfalls. By contrast, *Mateo*, who lives in a mountain town with plenty of access to water systems, observed that in spring of 2018, there was almost no rain. *Luis* laments that his town in the mountains used to be rich in water compared to now. Thus, spatial characteristics could also be affecting perceptions of the intensity of such changes.

Climate change concern

Narratives around concerns about climate change revealed connections to *relational cultural spaces* at the global, local, and intergenerational levels. When I asked participants about their concerns in relation to climate change, they generally expressed high concerns about accelerated climate change regardless of their location. Some participants mentioned the global dynamics of environmental change, referring to deforestation in the Amazon and international climate agreements, for example. Others expressed concern about the normalization of climate change, for example, *Ángel* noted how he was worried that we are now getting used to these changes. Some also talked about the need to invest in renewable sources of energy to reduce greenhouse emissions. However, most expressed concerns regarding local and interpersonal matters, such as concern about the toll that climate change could take on their younger family members.

Place was also sometimes used to trivialize climate change by noting that climatic changes have always existed, i.e. referring to a point in the past at which similar weather changes where noticed in their town. Meanwhile, these same participants acknowledged the acceleration of these changes, or affirm that we should listen to scientists. Thus, knowledge systems linked to space can sometimes lead to misconceptions around the causes of climate change that do not correspond with scientific notions.

Finally, participants also expressed general environmental concern around constructions of space. Again, water was an important element, as several participants claimed to have seen careless and abusive handling of water supplies. Some also explained that they take actions to reduce water waste. Other participants admitted psychological or emotional pain induced by the sight of a dry stream or forest fire. They referred to the interactions of elements in ecosystems and the broader consequences of human activities that harm isolated ecosystem components, e.g. herbicides and draining water resources.

The analysis of climate change perception in depopulated communities revealed that participants were aware and concerned about climate change. Their climate change perception is also relatively consistent with the scientific literature. Furthermore, place is often used as a point of reference to conceptualize the effects of climate change, including experiences with loss.

Adaptation practices in depopulated communities

The interviews revealed fractures between the official and local narratives of climate change adaptation as well as within the local climate culture. The national and regional governments have plans and strategies for climate change adaptation. In fact, the region of Granada launched a provincial plan for climate adaptation in 2019. This plan contains an assessment of climate threats and vulnerabilities in the region as well as measures for action including agricultural practices, water management, and a monitoring and evaluation plan. However, when I asked participants about adaptation practices to climatic changes, several had a hard time thinking of specific measures adopted at the community or individual level. Thus, the seemingly coherent official climate culture was not present in the interviews in which *knowledge systems* around adaptation practices tended to be absent or presented as reactions to changes in spatial dimensions.

Commonly noticed adaptation practices were reactive as they were connected to responses against soil and weather changes in farmlands. Participants noted two main practices. First, behavioral changes in crop and soil management practices such as stopping the use of pesticides, no tillage, or introduction of bees. In addition, participants mentioned crop variety changes. Earlier springs prompt earlier blooms that in turn are more vulnerable to frost. In response to these impacts, farmers reported a move from native almond trees to a type that blooms later in the year. Other participants mentioned the incorporation of tropical crops in areas that did not used to have subtropical conditions. One of the interviewees characterized this change as 'unthinkable' a few years ago.

Second, participants noted water management changes, specifically the construction of wells and shifts to drip irrigation as a response to the lack of reliable rainfall as a means to increase crop production. These changes provoke conflicting reactions. On the one hand, some participants view the installation of drip irrigation, a localized irrigation system, as a positive step toward reducing water usage, i.e. moving away from surface irrigation. For example, *Luis* explains that: "40 years ago, we mostly used surface irrigation. Now almost everything is through drip irrigation, because there is less water and we need to make the most of it".

On the other hand, drip irrigation is associated with an increase in well drilling (Sese-Minguez et al. 2017) and, therefore, considered as maladaptation. Participants noted that irrigation modernization was particularly prominent on larger farms and in places that have access to underground water. For some farmers, most notably those in dryland areas, the increasing extraction of underground water interrupted the natural flow of water. Some such as Eduardo noted the incapacity to adopt the new irrigation systems:

> We are in a dryland area. We can't adapt to the new (…) there are fields that use to be dryland and now are irrigated but not every area can be converted, because we have micro-farms (…). Here the land is very distributed, so we can't adapt or change our olive trees to irrigation (…). When there is a period of drought (…) we have a hard time.

The tensions over water are sometimes manifested in conflicts. For example, *David* explains that when there is drought, he feels stressed and worried, and sometimes there might be fights with neighbors about water usage.

These narratives of adaption reflect fractures between official statements with some degree of coherence and local perceptions, which tend to reflect reactive rather than anticipatory adaptation practices or nonexistent adaptation policies. Similarly, contradictory perceptions of adaptation practices, e.g. sustainable vs. extractive, denote climate fractures.

Discussion and conclusion

Conflicting place identities emerge in rural depopulated areas due to various transitions and disturbances of spaces. This chapter has shown how the social costs of industrialization – that is, models of economic exchange – have fractured space, provoking deterritorialization, loss, and existential struggles. At the same time, traditional place identities persist, oscillating between notions of obsolescence such as ageing populations; and notions of sustainability such as traditional agriculture that conserves biodiversity. These conflicting place identities construct a fractured climate culture with inconsistent climate change adaptation practices.

A number of implications derive from this study. First, it is necessary to combine notions of the relational cultural approach with political ecology to fully understand climate cultures in depopulated rural Spain. Participants disclosed a high attachment to both the built and natural environment and used place as a point of reference to express knowledge systems around climate change. Indeed, place, its shape, and elements, most notably water, was used to articulate relatively consistent notions of climate change. At the same time, the study also demonstrates that notions of political ecology are relevant as marginalization and conflict over access to resources can generate a fractured cultural identity, and by extension, climate cultures.

Second, fractures are noted between the official and local climate cultures. While at the regional and institutional level the discourse on climate change seems to enjoy some stability, the local level conveys a greater degree of incoherence. On the one hand, locals communicated a relatively high degree of climate change and environmental awareness and concern. On the other hand, adaptation practices to climate change were sometimes described as reactionary or nonexistent. Furthermore, the adaptation practices discussed revealed conflicting attitudes, e.g. modernization of water systems was viewed as simultaneously promoting water resource conservation and increasing water system overuse.

The fractured climate culture in the studied areas is also explained by the shifts that rural depopulated areas are experiencing. In this way, the study builds from the concept of shadow landscapes to provide a precise description of the process of marginality in these areas. Indeed, shadow landscapes are not described as remote slowly changing units. On the contrary, in the studied towns, socio-economic and ecological forces, such as population decline, globalization, mechanization, and intensification procedures that respond to market demands and accelerated environmental changes, are constantly shaping spaces and social relations by creating deagrarianizated and deterritorialized places. First, a sense of loss of ecosystems, culture, employment, services, population, and space prevails in these areas. Second, farming, which depends on external factors such as subsidies and urban demands, tends to be presented as an agonizing endeavor. Environmental and economic risks, demanding working conditions, and job insecurity shape the everyday lives of farmers who lament the lack of value placed on their occupational identity, i.e. low market value of products and instable and low salaries.

This study does not come without limitations. First, the sample is mostly male, a product of a sampling strategy that focus on agricultural workers in areas with male overrepresentation. The voices and perspectives of other gender identities should also be featured in future research. Second, there are other relevant socio-natural dynamics that need further attention such as ethnic relations or local politics dynamics.

Despite these limitations, the analysis of shadow landscapes uncovers contradictions that might produce a fractured climate culture. This incoherence can prevent such communities from adopting solid climate change adaptation and mitigation practices in spaces that are already affected by changing climatic and social systems. It seems hard to construct a credible climate change imaginary to protect future generations in places that are losing their cultural traditions while also being threatened by marginality and lack of generational replacement. Special attention should be paid to communities that practice dryland and small-scale agriculture as these are more likely to be less resource extractive and tend to be more affected by water scarcity. All in all, the communities' overall high attachment to place and acute environmental awareness present an opportunity for the cultivation of climate protection practices. In addition, existing channels to preserve and promote traditional and sustainable agroecological practices present an opportunity to implement mitigation and adaptive responses to climate change such as regenerative agriculture.

Acknowledgments

This study received funding from the Center for European Studies at the University of Florida. I would also like to thank all participants of the study for sharing their input and stories with me.

References

Adger, Neil. 2016. "Place, Well-Being, and Fairness Shape Priorities for Adaptation to Climate Change". *Global Environmental Change* 38: 1–3.

Brace, Catherine, and Hilary Geoghegan. 2011. "Human Geographies of Climate Change: Landscape, Temporality, and Lay Knowledges". *Progress in Human Geography* 35 (3): 284–302.

Bryant, Raymond, Ángel Paniagua, and Thanasis Kizos. 2015 "Governing People in Depopulated Areas". In *The International Handbook of Political Ecology*, edited by Raymond L. Bryant, 233–245. Cheltenham (United Kingdom): Edward Elgar Publishing.

Cabello, Javier, and Antonio Castro. 2012. *Estado y Tendencia de Los Servicios de Los Ecosistemas de Zonas Áridas de Andalucía* [Status and Trend of Ecosystem Services in Arid Zones of Andalusia]. https://www. juntadeandalucia. es/.../portal.../ema_aridos.pdf

Coll, Joan Ramon, Philip Jones, and Enric Aguilar. 2015. "Expected Changes in Mean Seasonal Precipitation and Temperature across the Iberian Peninsula for the 21st Century". *Quarterly Journal of the Hungarian Meteorological Service* 119 (1): 1–21.

de Castro, Manuel, Javier Martín-Vide, and Sergio Alonso. 2004. "El Clima de España: Pasado, Presente y Escenarios de Clima para el Siglo XXI" [The Climate of Spain: Past, Present and 21st Century Climate Scenarios]. In *Evaluación Preliminar de los Impactos en España por Efecto del Cambio Climático* [Preliminary Assessment of the Impacts of Climate Change in Spain], edited by Ministerio de Medio Ambiente, 1–64. Madrid (Spain): Ministerio de Medio Ambiente. https://www.miteco.gob.es/es/cambio-climatico/temas/impactos-vulnerabilidad-y-adaptacion/evaluacion_preliminar_impactos_completo_2_tcm30-178491.pdf del Pino, Julio, and Luis Camarero. 2017. "Despoblamiento Rural: Imaginario y Realidades" [Rural Depopulation: Imaginary and Reality]. *Soberania Alimentaria, Biodeiversidad y Culturas* 27: 6–10.

Devine-Wright, Patrick. 2013. "Think Global, Act Local? The Relevance of Place Attachments and Place Identities in a Climate Changed World". *Global Environmental Change* 23 (1): 61–69.

Elshof, Hans, and Ajay Bailey. 2015. "The Role of Responses to Experiences of Rural Population Decline in the Social Capital of Families". *Journal of Rural and Community Development* 10 (1): 72–93.

Entrena Durán, Francisco. 1999. "La Desterritorialización de Las Comunidades Locales Rurales y Su Creciente Consideración Como Unidades de Desarrollo" [The Deterritorialization of Rural Communities and Their Growing Consideration as Units of Development]. *Revista de Cooperativismo Agrario y Desarrollo Rural* 3: 29–41.

Entrena Durán, Francisco. 2010. "Dinámicas de los territorios locales en las presentes circunstancias de la globalización" [Dynamics of Local Territories in the Present Circumstances of Globalization]. *Estudios sociológicos*, 28 (84): 691–728.

Espín Piñar, Rocío, Eduardo Ortiz Moreno, and José Ramon Guzmán Álvarez. 2010. *Manual del Acequiero* [Oiler's Manual]. http://www.juntadeandalucia. es/medioambiente/portal_web/agencia_andaluza_del_agua/participacion/ publicaciones/manual_del_acequiero.pdf

European Commission. 2014. "Community-led Local Development". https://ec. europa.eu/regional_policy/sources/docgener/informat/2014/community_en.pdf

García-Mozo, H., L. Yaezel, J. Oteros, and C. Galán. 2014. "Statistical Approach to the Analysis of Olive Long-Term Pollen Season Trends in Southern Spain". *Science of the Total Environment* 474: 103–109.

Gieryn, Thomas F. 2000. "A Space for Place in Sociology". *Annual Review of Sociology* 26 (1): 463–496.

Gómez-Baggethun, Erik, Victoria Reyes-Garcia, Per Olsson, and Carlos Montes. 2012. "Traditional Ecological Knowledge and Community Resilience to Environmental Extremes: A Case-Study in Doñana, SW Spain". *Global Environmental Change* 22 (3): 640–650.

González de Molina, Manuel, David Soto-Fernández, Gloria Guzmán-Casado, Juan Infante-Amate, Eduardo Aguilera-Fernández, Jaime Vila-Traver, and Roberto García-Ruiz. 2019. *The Social Metabolism of Spanish Agriculture, 1900–2008: The Mediterranean Way towards Industrialization*. Cham (Switzerland): Springer Nature.

Heimann, Thorsten. 2018. *Culture, Space and Climate Change: Vulnerability and Resilience in European Coastal Areas*. London (United Kingdom): Routledge.

INE (National Institute of Statistics). 2019. "Contabilidad Regional de España" [Regional Accounts of Spain]. https://www.ine.es/prensa/cre_2018_2.pdf

INE. 2020. "Cifras Oficiales de la Población de los Municipios Españoles" [Official Population Figures for the Spanish Municipalities]. https://www.ine.es/dynt3/ inebase/es/index.htm?padre=517&capsel=525

Infante-Amate, Juan, Inmaculada Villa, Felipe Jiménez, Manuel M. Martín, David M. López, Geoff Cunfer, and Manuel González de Molina. 2016. "The Rise and Fall of the Cortijo System: Scattered Rural Settlements and the Colonization of Land in Spain's Mediterranean Mountains since 1581". *Journal of Historical Geography* 54: 63–75.

Junta de Andalucia. 2020. "Argos". http://www.juntadeandalucia.es/servicioandaluzdeempleo/web/argos/web/es/ARGOS/index.html

Junta de Andalucía. ND. "Desarrollo Rural" [Rural Development]. https://www.juntadeandalucia.es/organismos/agriculturaganaderiapescaydesarrollosostenible/areas/desarrollo-rural/veinte-anos.html

Madejón, P., C. Barba-Brioso, N. W. Lepp, and J. C. Fernández-Caliani. 2011. "Traditional Agricultural Practices Enable Sustainable Remediation of Highly Polluted Soils in Southern Spain for Cultivation of Food Crops". *Journal of Environmental Management* 92 (7): 1828–1836.

Marshal, Nadine, Sarah Park, W. Adger, K. Brown, and Stuart M. Howden. 2012. "Transformational Capacity and the Influence of Place and Identity". *Environmental Research Letters* 7 (3): 1–9.

Masterson, Vanessa A., Richard Stedman, Johan Enqvist, Maria Tengö, Matteo Giusti, Darin Wahl, and Uno Svedin. 2017. "The Contribution of Sense of Place to Social-Ecological Systems Research: A Review and Research Agenda". *Ecology and Society* 22 (1): 49.

Ortiz-Miranda, Eladio, Vicente Arnalte-Alegre, and Ana Maria Moragues-Faus. 2013. *Agriculture in Mediterranean Europe: Between Old and New Paradigms.* Bingley (United Kingdom): Emerald Group Publishing.

Paniagua, L. L., A. García-Martin, F. J. Moral, and F. J. Rebollo. 2019. "Aridity in the Iberian Peninsula (1960–2017): Distribution, Tendencies, and Changes". *Theoretical and Applied Climatology* 138 (1–2): 811–830.

Peña-Gallardo, M., S. R. Gamiz-Fortis, Y. Castro-Diez, and M. J. Esteban-Parra. 2016. "Comparative Analysis of Drought Indices in Andalusia during the Period 1901–2012". *Cuadernos de Investigacion Geografica* 42 (1): 67–88.

Pinilla, Vicente, and Luis Sáez. 2017. *La Despoblación Rural En España: Génesis de Un Problema y Políticas Innovadoras* [Rural Depopulation in Spain: The Genesis of a Problem and Innovative Policies]. Zaragoza (Spain): Ceddar.

Sese-Minguez, Saioa, Harm Boesveld, Sabina Asins-Velis, Saskia Van der Kooij, and Jerry Maroulis. 2017. "Transformations Accompanying a Shift from Surface to Drip Irrigation in the Cànyoles Watershed, Valencia, Spain". *Water Alternatives* 10 (1): 81–89.

Sinoga-Ruiz, José, Ramon-Garcia Marín, Juan Francisco Martínez Murillo, and Miguel Ángel Gabarrón Galeote. 2011. "Precipitation Dynamics in Southern Spain: Trends and Cycles". *International Journal of Climatology* 31 (15): 2281–2289.

Suárez-Navaz, Liliana. 2004. *Rebordering the Mediterranean: Boundaries and Citizenship in Southern Europe.* New York: Berghahn Books.

Tschakert, P., N. R. Ellis, C. Anderson, A. Kelly, and J. Obeng. 2019. "One Thousand Ways to Experience Loss: A Systematic Analysis of Climate-Related Intangible Harm from around the World". *Global Environmental Change* 55: 58–72.

UPA (Union of Small Farmers). 2020. "Las 10 principales reivindicaciones de los agricultores y ganaderos al límite" [Top 10 Demands of Farmers and Ranchers at the Limit]. https://www.upa.es/upa/noticias-upa/2020/3088/

Wainwright, Megan, and Andrew Russell. 2010. "Using NVivo Audio-Coding: Practical, Sensorial and Epistemological Considerations". *Social Research Update* 60: 1–4.

6 Cultural Perception and Adaptation to Climate Change among Reindeer Saami Communities in Finland

Klemetti Näkkäläjärvi, Suvi Juntunen and Jouni J. K. Jaakkola

Introduction

The Arctic region faces particularly severe effects of climate change (IPCC 2014). In Finland, the average temperature has risen 2.3°C since the pre-industrial period (Mikkonen et al. 2015). If warming proceeds at this rate, future climatic conditions in Northern Finland will resemble those currently prevailing in Central Finland (Ruosteenoja 2016). Many arctic indigenous people have already observed widespread effects of climate change by the end of the 20th century (Archer et al. 2017; Moerlein and Carothers 2012). Swedish Saami first observed signs of climate change in the 1970s, and in recent years, these changes have been accelerating (Furberg et al. 2011).

In previous studies, the effects of climate change on reindeer herding are evaluated in terms of livelihoods, the market economy, biology, the environment, and adaptation, as well as reindeer herding as a tool for climate change mitigation (Eira et al. 2018; Käyhkö and Hortskotte 2017; Cohen et al. 2013; Furberg et al. 2011). More recently, researchers have examined the holistic effects of climate change on Saami people (Jaakkola et al. 2018).

This chapter presents some of the findings of the SAAMI – Adaptation of Saami People to Climate Change – Project implemented in 2019–2020. This chapter provides new theoretical insights to better understand environmental perception and climate adaptation in Reindeer Saami culture. The specific methodology and other findings are reported elsewhere (Näkkälä-järvi et al. 2020). The main objective of the project was to produce holistic information on climate change and Saami culture for decision-makers and Saami communities themselves.

The specific goal of this chapter is to use ethnographic data to analyze, first, how Saami reindeer herders perceive the environment and effects of climate change and, second, how Reindeer Saami communities have adapted to the changing climate. The study spans the period from 1960 to 2018. The chapter examines how socially shared climate change knowledge (Heimann 2019, 59) is formed and analyzed in the Reindeer Saami context. We identify this knowledge as "landscape memory" that is formed and developed spatially in an enculturation processes.

DOI: 10.4324/9781003307006-10

Previous research and theoretical background

The concept of culture is highly versatile; it can be understood as a way of life of the community, including spiritual, material, environmental, and social aspects of culture (Herskovits 1958). In the framework of climate cultures, the concept of culture can be restricted to aspects of socially shared knowledge related to climate change, e.g. understandings of climate change, shared vulnerability perception, and/or adaptation practices (Heimann 2019, 19). The broad concept of climate cultures in this chapter refers to cultural understandings of climate change and climate perception among Reindeer Saami communities in Finland.

Anthropology plays a key role in understanding and conceptualizing the socio-cultural effects of climate change and climate adaptation (Fiske et al. 2014; Crate 2011; Pelling 2011). Ethnography studies cultural phenomena and communities from the points of view of research subjects (emic perspective). The emic or internal perspective represents the way in which members of a culture perceive, structure, and value matters within their own sphere of life. Ethnography refers to both the actual field work and the analysis. The role of ethnography in climate change research is to elucidate climate change in a cultural and social context, as well as examine opportunities for communities to adapt and act (Fiske et al. 2014, 19, 61). With ethnography, a researcher can explore climate change as a cultural phenomenon in the context of social life (see Fiske et al. 2014, 21).

Our methodology can be conceptualized as "climate ethnography", a concept introduced by Crate (2011). Climate ethnography entails the development of a critical, collaborative ethnography that integrates human perceptions and understandings and further develops cultural models (Näkkäläjärvi and Juntunen 2022, forthcoming). We have aimed to approach the topic through the Saami culture's own concepts and traditions, using the Saami language and cultural classification system, and collaborating with the Saami communities and Saami stakeholders. Local observations can make an important contribution to understanding the pervasive impact of climate change on ecosystems and societies (Savo et al. 2016).

Ethnography plays a key role in producing new knowledge from the perspective of people coming from different cultural backgrounds and ethnicities. We argue that ethnographic research is becoming increasingly important in the global world, as small languages and cultures are becoming increasingly endangered because of climatic, environmental, and societal processes. Clifford criticizes this kind of ethnography as "salvage ethnography," where the ethnographer seems to "save" vanishing cultures while reminding us that almost all cultures have already been textualized (Clifford 1986, 112–113, 117). Although Saami society is a literate society, much of its knowledge is still in oral form, especially regarding cultural perception of the environment. Ethnographic data and analysis on climate perception among Reindeer Saami in Finland indicate that environmental knowledge, livelihood, and cultural traditions are developing but also that, simultaneously, traditions and knowledge are disappearing. Climate ethnography can provide new knowledge on climate perception and adaptation and help to "textualize" and maintain traditions and knowledge for future generations.

Indigenous peoples' knowledge related to the perception of the environment and climate has been commonly described as traditional knowledge, traditional ecological knowledge, or local knowledge (see Eira et al. 2018; Riedlinger and Berkes 2001). This knowledge is not generally considered as science, although the concept is widely used in political (Convention of Biological Diversity 1992, article 8 (j)) and scientific literature. The IPCC has described traditional knowledge as a valued knowledge system that can, together with or independently of natural sciences, and produce useful knowledge for climate change detection and adaptation (IPCC 2014, 1001).

Methodology

The Saami are indigenous people that inhabit *Sápmi* (land of the Saami in the North Saami). It covers areas in Northern Finland, Sweden, Norway, and the Kola Peninsula. There are around 11,000 Saami in Finland, with over 60% living outside of the traditional Saami home region (see Figure 6.1). There are three Saami languages in Finland, the North, Inari, and

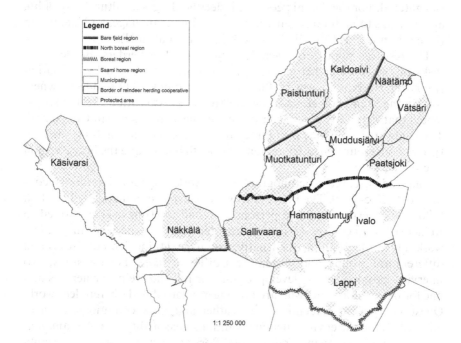

Figure 6.1 Map of Saami home region and reindeer herding cooperatives.
Source: Reindeer Herder's Association, Metsähallitus, National Land Survey of Finland.

Skolt Saami, all of which are considered endangered. Less than half of the indigenous Saami speak Saami as their native tongue.

The major competing land uses in the study region are forestry, tourism, and mechanical gold placer mining. Traditional Saami livelihoods are reindeer herding, fishing, hunting, gathering, and handicrafts (*Sámi duodji*). The Saami have faced profound socio-economic and cultural changes that have impacted their languages, society, income, habitat, livelihoods, customs, and sense of community. Saami culture is considered endangered (Jaakkola et al. 2018) and therefore, the starting points for adapting to climate change are challenging. The endangerment of Saami culture has been influenced by the state's assimilation policy, economic transformation, and the legal and public status of the Saami.

The traditional Saami structure to manage reindeer work and pasture areas is a *siida*. It is a kinship-based Saami communal structure that takes into consideration pasture conditions, labor needs, and other pressures (Näkkäläjärvi 2002). Reindeer work includes, among other things, herding, maintenance of social relations, calf marking, separations, slaughtering, and supplementary feeding (where available), maintenance of herding equipment and infrastructure, meat trade, and administration. The separation of reindeer takes place in autumn or winter. A small number of reindeers are herded into a corral at a time. The herder selects the reindeer to be left alive, the ones that will be sold to meat buyers, and the ones that will be slaughtered. For the Saami people, reindeer herding is a cultural way of life and a means to maintain their traditions, culture, language, and identity in the modern world (Näkkäläjärvi 2013; Furberg et al. 2011).

The research population includes Saami reindeer herders. Borders and tasks of the reindeer herding cooperatives are determined by legislation (Reindeer Husbandry Act 1990). There are around 1,220 reindeer owners total in the Saami homeland (Reindeer Herder's Association 2019). It is estimated that 70%–80% of reindeer owners in the Saami homeland are Saami. The study area is divided into three different areas in the analysis: the bare fjeld region (mountain or fell region), the north boreal, and the boreal region (see Figure 6.1).

The study is based on ethnographic fieldwork, including interviews, conducted in 2015–2019. The informants were selected systematically based on following criteria: they were persons who had to (1) have enculturated to Reindeer Saami culture, (2) have at least ten years of experience in reindeer work, (3) speak Saami as their native language, (4) include both sexes and different age groups, and (5) represent different *siidas*. For the study, two interview protocols were developed, one for full-time reindeer herders and one for older informants who are no longer involved in daily reindeer work. Questions covered perceptions of weather and snow conditions, changes and stability in the environmental conditions, seasonality, use of Saami language, issues of safety in reindeer work, flora and fauna, as well as causality and reindeer work during the time span of the study. Of 72 total informants,

21 were women (29%) and 51 were men (71%). The average age of informants was 61 years, ranging from 21 to 84 years. Of those interviewed, 54 reported being active in reindeer work and 18 reported being retired from reindeer work. Among the interviewees, 60 identified as belonging primarily to North Saami, 6 to Inari Saami, and 6 to Skolt Saami lingual and cultural groups.

Empirical results

Reindeer work models

In the research literature, reindeer herding has been commonly considered uniform throughout the Sápmi region. Regional and cultural differences in reindeer work models have been overlooked (see Pekkarinen et al. 2015; Furberg et al. 2011). Saami reindeer herding culture throughout the whole Sápmi shares several common features and cultural backgrounds, such as the *siida* system, language, cultural identity, pasture cycle, landscape memory, and a kinship-based reindeer earmark system (*mearkaoalli*). The earmarks of family members resemble each other and are similar to, or a variation of, earmarks of previous generations (Näkkäläjärvi 2013). We have identified ten different reindeer work models in our study area (see Table 6.1).

Reindeer work models refer to the ways in which reindeer herding as a livelihood is carried out. Reindeer work is a translation of the North Saami concept *boazobargu*. In this context, the words "reindeer work" and "reindeer work model" are used because these concepts best correspond to the informants' own understanding of their livelihood and culture. The identified reindeer work models reflect adaptation to geography, environmental conditions, local traditions and communities, innovations, the linguistic environment, governance, competing land uses, the economy, and climate change.

The identified reindeer work models play a critical role in understanding new cultural models and ethnographic reality in a changing climate (see Crate 2011). Communities adapt to climate change in different ways, even though the starting points are the same. All Saami reindeer herding work models are based on the common work model (see Figure 6.2), but different communities have chosen different ways to adapt. We argue that adaptation to climate change can increase cultural diversity, and this also requires changes in the livelihood subsidy system and administration.

There are variations within the models that generalized definitions cannot bring to light. All the different reindeer work models require specific know-how, skill, and terminology, whereby some of the knowledge and skills needed in one model are unnecessary and disappear in other models. It is important to recognize the variability of work models if one is to understand different adaptation processes within Reindeer Saami culture. There may be more than one reindeer work model in one cooperative and some of the models may be combinations of other models.

Table 6.1 Identified reindeer work models, status, and introduced adaptation measures

Reindeer work model	Description	Start of adaptation measures	Introduced adaptation measure	Current state
Traditional reindeer nomadism (1)	Reindeer work is managed by siida. Supplementary feeding is not used. Technology such as GPS-collars can be used occasionally, but technology has not replaced traditional navigational and reindeer identification skills. Calves are caught with suohpan (lasso) in summer separations for earmarking.	1990	Increase in herding; strengthening the siida system; change of pasture cycle	Endangered, prosperous locally
Reindeer nomadism that utilizes supplementary feeding (2)	Reindeer work is based on siida system. Reindeers are fed with supplementary feeding in winter and spring. Calves are marked in calving enclosures or in autumn in separation corrals. Reindeers are caught by hand.	1990	Strengthening the siida system. Supplementary feeding, permanent transformation of reindeer work model; use of calving corrals, change of pasture cycle.	Prosperous throughout the Saami home region
Combined reindeer nomadism (3)	Reindeer work is based on siida system. Reindeers are fed with supplementary feeding in winter and spring. After winter separations reindeer herders take part of the herd to corrals that are located near the houses. Calves are marked in summer and caught with lasso.	1990	Supplementary feeding, permanent transformation of reindeer work model, change of pasture cycle	Rare
Adaptive reindeer nomadism (4)	Feeding is for herding purposes during winter with hay made by the herders themselves and with dried bunches of birch branches. Small amounts of store-bought pellets can be used. Siida system is central. Calves are marked in summer and caught with vimpa (a stick that has a lasso on top of it) or by hand.	2000	Strengthening of the siida system	Prosperous locally

Reindeer work of managed pasture cycle (5)	The grazing area is divided by fences to seasonal pastures. Supplementary feeding can be used. Calves are caught with vimpa, by hand or calves are marked in calving enclosures.	2000	Increase in herding, supplementary feeding, division of pasture land	Prosperous
Innovation-oriented reindeer work (6)	Reindeers are fed with supplementary feeding in winter and spring. Technological innovations: GPS-collars, helicopters, and drones are used to track and herd the reindeer. Female reindeers are commonly kept for calving in corrals where the calves are earmarked. Calves are caught by hand or with vimpa.	2000	Introduction of new technology, supplementary feeding, permanent transformation of reindeer work model, change of pasture cycle	Prosperous
Reindeer work managed by the reindeer cooperative (7)	Cooperative is responsible for organizing reindeer work. Herders are paid various compensations by the cooperative. Calves are caught by hand or with vimpa.	1990	Increase in herding, change of pasture cycle	Prosperous
Reindeer work of combined livelihoods (8)	A reindeer work model in which fishing is an essential part. Calves are marked in autumn separations. The calves are caught by hand. Supplementary feeding is used, mainly hay.	1990	Supplementary feeding; permanent transformation of reindeer work model; change of pasture cycle	Prosperous
Tourism oriented reindeer herding (9)	A reindeer work model in which the provision of tourism services is an important part of a reindeer herder's economy and livelihood. Tourism reindeers are reared and fed part of the year. Calves are caught with a vimpa or in the calving corral.	1990	Supplementary feeding, introduction of tourism services, permanent transformation of reindeer work model	Prosperous
Reindeer racing as part of reindeer work (10)	A reindeer work model in which reindeers are bred and trained into reindeer racing. Racing reindeers are kept in corrals and fed with fodder. Rest of the reindeer are herded according to siida/cooperative practices.	—	No specific adaptation measures, adaptation measures are implemented according to siida/cooperative practices	Prosperous

Source: Own representation.

Figure 6.2 illustrates the timeline of when different models have evolved during the study period (1960–2018) from traditional reindeer nomadism to different reindeer work models. The analysis in Figure 6.2 reflects the historical knowledge of our informants. The differentiation of reindeer work models already began in the 1980s.

Perception of climate change and its impacts on
Reindeer Saami knowledge

The environmental and livelihood knowledge of Saami is a systematic knowledge system that can be identified as landscape memory theory (Näkkäläjärvi 2013, 43–44). Culturally developed landscape memory can be considered as the core of Reindeer Saami culture. It is based on the spatial relationship between the individual and his/her ethno-ecological niche. It is formed in the enculturation process, meaning the process by which an individual learns the knowledge and skills relevant for reindeer work and Reindeer Saami culture. Through landscape memory, herders identify the linguistic, symbolic, material, and functional dimensions of the cultural landscape and their relationship with the environment and their reindeers (see Figure 6.3). Throughout this process, they learn to categorize phenomena, as well as navigate and function in the landscape for reindeer work – in other words, they learn to see the landscape culturally (Näkkäläjärvi 2013, 29 f. 42 ff.).

Landscape memory is a knowledge system, a tool for perceiving and adapting to the changes in the environment, and a causal classification and information system. Herders observe the changes in the environment in relation to reindeer, their nutritional needs and conditions, seasons, snow and weather conditions, available resources, and causal relationships, and they make decisions while taking into consideration future needs for pasture use, i.e. the material and functional dimensions of the landscape. The knowledge of herders has been, in some studies, identified as a matter of intuition (Ingold 2000). Lavrillier and Gabyshev (2018) use the concept "emic sense of climate" to understand and analyze the Evenki environmental knowledge system, arguing that the Evenki knowledge on the environment is a theoretical knowledge system, and within the system, they produce hypotheses focusing on the interactions between the climate, landscape, flora, and fauna. Our findings indicate that the Saami concept of landscape memory as a cultural knowledge system has similarities with the environmental observation systems of other indigenous peoples, although there are conceptual differences in definitions.

> When I move in the terrain, I look at the different formations, conditions, vegetation, animals, reindeer, people, and traces in the landscape; I feel the wind and hear the sound. I always know where I am, I remember the places under all conditions. In my mind, all the places and conditions are in the Saami language. I know the mountains by their form,

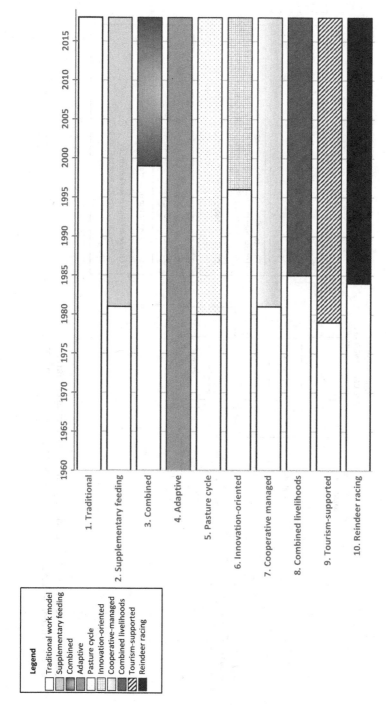

Figure 6.2 Development of reindeer work models in the research area 1960–2018.
Source: Own representation (see also Näkkäläjärvi et al. 2020, 59).

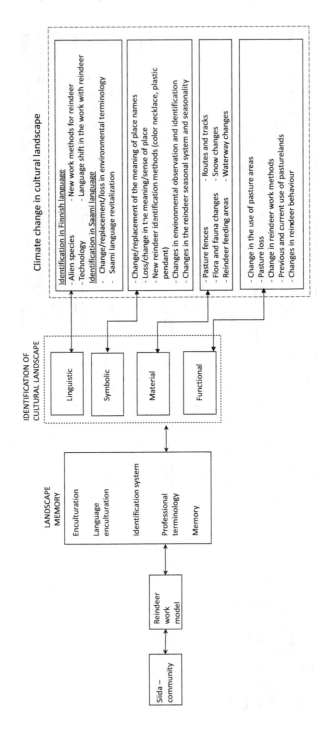

Figure 6.3 Perception of the environment in a changing climate.
Source: Own representation (adaptation from Näkkäläjärvi et al. 2020, 276).

in their relation to other mountains, and by location and place name. It is easy to track reindeer in snow, because we have so many words for different tracks.

(Informant 1)

Herders do not use GPS or maps for navigation. Herder's daily pasture routes can vary from 20 to 200 kilometers. They must know the landscape in different seasons and light conditions, in extreme conditions, and adapt to the changed landscape in areas used for forestry. Place formations, toponyms, knowledge of direction, and engrams from routes and landscape create elaborative focal points for memory to help herders navigate and function in the landscape and monitor environmental changes (compare with Psathas 1972).

The SAAMI-project has classified Saami climate change observations into 80 different categories. Saami have observed all changes in detail and studied the causality of changes. Some of the observations were not known to academic sciences, such as causal effects of defoliation by geometrid moths resulting in the disappearance of mushrooms (Näkkäläjärvi et al. 2020). The defoliation has also brought positive effects, for instance, lichen (*lichenas*) has spread to the affected area. Lichen does not thrive in birch forests. Climate-induced changes can have both positive and negative effects at the same time. Mushrooms are vital for the reindeer to build up fat reserves for mating season and winter.

The incidence of defoliation by geometrid moths has increased. The cycle is shortened, and the birches don't have enough time to recover. The defoliation influences reindeer because they feed on the birch leaves. After defoliation, the mushrooms disappear. In bad mushroom years, winter survival (of reindeer) is worse.

(Informant 2)

Informants stressed that the climate in the Saami homeland has always been variable, though predictable. Overall, the interviewees have clearly articulated their experiences with the difference between normal variations in weather conditions and climate change. They have also described permanent climate changes. This creates additional challenges to their livelihoods, as fluctuations of temperature and extreme weather conditions reduce predictability in herding. Table 6.2 classifies, at a general level, the observed changes in the environment. Observations are influenced by the age of the informants, older informants date the first signs of climate change to the 1980s or earlier, and younger informants date the first signs to the 1990s and 2000s. Saami observations are in line with those of other Arctic indigenous peoples (Archer et al. 2017; Moerlein and Carothers 2012). The changes have accelerated in the 21st century. In particular, the occurence of extreme weather events have accelerated.

Table 6.2 First signs of climate change

Observations of changes in the environment	Area	1960	1980	1990	2000	2010
Changes in vegetation	Bare fjeld region		First observation			
	Northboreal		First observation			
	Boreal		First observation			
Changes in wind (strength/velocity)	Bare fjeld region			First observation		
	Northboreal				First observation	
	Boreal				First observation	
Ever-changing weather conditions	Bare fjeld region		First observation			
	Northboreal			First observation		
	Boreal	First observation				
Changes in snow (amount/ quality/precipitation)	Bare fjeld region		First observation			
	Northboreal				First observation	
	Boreal				First observation	
Changes in ice (carrying capacity, melting)	Bare fjeld region		First observation			
	Northboreal				First observation	
	Boreal				First observation	
Changes in mean temperature & fluctuation	Bare fjeld region			First observation		
	Northboreal				First observation	
	Boreal				First observation	
Changes in seasonality	Bare fjeld region			First observation		
	Northboreal				First observation	
	Boreal				First observation	
Changes in animal species	Bare fjeld region				First observation	
	Northboreal				First observation	
	Boreal				First observation	
Changes in *guohtun* (pasture & nutrition conditions)	Bare fjeld region		First observation			
	Northboreal		First observation			
	Boreal				First observation	

As the region warms, forest area and shrub growth has accelerated in all study regions. The greening process makes it difficult to navigate, travel, identify, search for, and herd reindeer in the landscape during the thawed period. In the boreal and northern boreal regions, informants have reported that moss is replacing lichen, the main diet for reindeer in winter. Palsa mires have disappeared from the boreal region, and partially from the northern boreal region and the bare fjeld region. Palsa mires are important places for reindeer nutrition and they are the best places for picking cloudberries (*rubus chamaemorus*). Informants have reported increased predation and that sea eagles have started to prey. Informants inferred that increased predation was a result of combined effects of climate change and nature conservation.

> I look at snow only as guohtun (pasture and nutrition condition). Over ten years ago, we always knew the guohtun conditions already in autumn when the first lasting snow cover came. Now we know what kind of guohtun there will be only as late as in January. I classify guohtun in Saami language by snow depth, quality, ice crust, and the ability of reindeer to dig for lichen. The snow used to drift in a similar way in the same places every year, but not anymore. The strong wind and changes in snow quality make it drift differently.
>
> (Informant 3)

For the Saami, snow is not a noun for "(a) small, soft, white pieces (called flake) of frozen water that fall from the sky in cold weather; (b) this substance when it is lying on the ground" as the Oxford English dictionary defines it, but a cultural conceptual frame: snow, *muohta*, covers over 450 different exact scientific definitions of different snow structures, movements, and formations in North Saami language, including shared cultural knowledge on its effects on the movement of herder and reindeer, and on the nutritional supply of reindeer.

Herders have reported that the changes brought about by climate change have affected their occupational safety and increased physical and mental stress. Mental stress is stemming from multiple environmental, economic, and societal challenges, including climate change. All-terrain vehicles (ATVs or quad bikes) that are used for pasture work in snow-free time, and snowmobiles, have dropped through ice and fallen in the snow. There have been several, even fatal, accidents due to changes in environmental conditions. The increase in insecurity has been experienced among many Arctic indigenous people (see Prno et al. 2011).

> Conditions have changed, and the seasons have changed as well. Spring is snowy, but the conditions in autumn-winter are different; there is less snow. Our region is very rocky, and it is very dangerous to ride snowmobiles in rocky terrain when there is little snow.
>
> (Informant 4)

Climate change has brought additional economic costs, such as supplementary feeding (pellets, hay, and/or lichen), and has weakened the economy of reindeer herders. Reindeer deaths have increased due to severe winter conditions, affecting subsidy levels and meat sales negatively. Similar observations have been reported by Saami in Sweden (Furberg et al. 2011).

> Nowadays, it is common everywhere in Sápmi that the herders have only little to sell. In some areas, the reindeer are in poor condition. Researchers tell us that we are to blame for poor pasture conditions, but there is very little we can do. The winter separations are always late, because we must wait for the lakes and rivers to freeze before we can herd the reindeer for the separations. We have to keep enough reindeer in reserve to prepare for predation and hard winters; we have to look to the future in our work and not make decisions based only on today's situation.
>
> (Informant 5)

Overall, the effects of climate change on reindeer herding were considered largely negative by the informants. As vegetation and weather conditions have changed, reindeer work and reindeer grazing have become more difficult. Positive effects were expressed by reindeer herders from the *Kaldoaivi* and *Paistunturi* reindeer cooperatives (work model 8). Longer autumns and milder winters have made it easier for reindeer to find nutrition, and more nutrition is available. The carcass weights of reindeer have increased, due to the combined effect of climate change and supplementary feeding.

Supplementary feeding improves reindeer's chances of surviving the winter. Informants also report that the need for supplementary feeding in reindeer herding has arisen because increased state controls have reduced the flexibility of the reindeer herding system. These measures may increase the herders' vulnerability as they become more dependent on the state (Rees et al. 2008). Keskitalo (2008), on the contrary, has argued that the changes in reindeer herding indicate the high adaptive capacity of reindeer herders. It is argued that the reasons for supplementary feeding are poor pasture conditions and government subsidies (Pekkarinen et al. 2015). However, study informants have identified climate change, lack of herders, loss of pasture, and social reasons as the main reasons for it.

Intergenerational effects of climate change

Climate-induced changes have brought uncertainty to the region, as the changes cannot be conceptualized and foreseen based on traditionally shared landscape memory. The Saami have traditionally recognized eight seasons, each season having its own typical conditions and reindeer work tasks. Currently, there are only four seasons clearly recognized. Summer and autumn come earlier and last longer. The Saami have so-called "name days" and "name weeks" that describe environmental conditions and activities related to traditional livelihoods. These have been used to predict conditions and weather. Since the weather and climate conditions have changed, the symbolic knowledge of the traditional annual cycle is disappearing. Changes affect the Saami conception of time, as well as the structure of traditional seasonal tasks. Saami have a strong connection to their landscape; changes in environmental conditions have a direct impact on their identity and weaken the connection between people, landscape, and culture (see also Cunsolo Willox et al. 2012). For instance, informants reported linguistic effects of climate change. There is no term in Saami languages for new phenomena or invasive alien species, and some of the terms disappear from the vocabulary because the phenomenon or condition no longer occurs due to changed conditions: "one consequence of climate change is that old knowledge and skills will be lost" (Informant 6).

A major concern that emerges from several interviews is the loss of traditional knowledge because young people do not learn reindeer work-related knowledge and skills in a natural, experiential way (compare with Cunsolo Willox et al. 2012; Pearce et al. 2010). Thus, climate change induces divergence between generations regarding knowledge and skill. The knowledge of elder generations can be overlooked by younger generations who are better able to utilize new technologies in adaptation work. The herder's landscape memory develops as a response to new conditions. A new type of knowledge has emerged regarding the supplementary feeding of reindeer and implementation of new technology, such as using drones in reindeer herding, resulting in new reindeer work models. The use of supplementary feeding affects the identification knowledge and weakens the importance of *guohtun* knowledge. New knowledge has also come about regarding effects of different snow conditions on driving snowmobiles.

> Predicting safety conditions for reindeer herders is like playing Russian roulette. In our region, reindeer also graze on islands. Snow conditions, wind, and fluctuations in temperature affect the ice conditions and the carrying capacity of ice. We cannot be sure anymore when it is safe to cross the ice in autumn-winter or even in winter.
>
> (Informant 8)

As a result of climate change and adaptation, changes are taking place that also affect landscape memory. Changes in livelihood models and the environment have led to the loss and replacement of some skills and knowledge. The introduction of a new technology may serve as a means of adaptation, but at the same time, other knowledge may be lost. Climate change is generational in nature: knowledge and skills not used by the current generation will not be passed to future generations. Informants have also pointed out that if environmental conditions continue to deteriorate, the reindeer herding livelihood will provide employment for an even smaller number of Saami.

This can accelerate emigration out of the Saami home region. As the Saami population population of the community decreases, it will also become more difficult to carry out traditional livelihoods and transfer landscape memory.

This new uncertainty poses serious challenges to herders' knowledge, professional know-how, and ability to rely on their landscape memory; no longer can a herder always trust his/her knowledge, the knowledge that has accumulated during centuries and that has safeguarded the success of Saami reindeer culture under harsh conditions into modern times. Climate change creates a new material layer in the landscape: the landscape of risks, dangers, and losses. The intergenerational effects of climate change adaptation are significant. The model of reindeer work is transmitted from one generation to the next in Saami siidas, although formal education is available. The next generation of herders will acquire a landscape memory and adopt a reindeer work model already adapted to climate change.

In sum, central concerns for the future are whether the Saami languages will survive as the working languages in reindeer herding, and whether the cultural richness and systematic knowledge system embedded in the language will survive in a changing climate, serving as a tool of landscape memory and in climate adaptation. The knowledge accumulated in landscape memory, namely, reindeer routes, preferred places in different seasons, conditions, and even identification of reindeer, becomes less important when using new technologies. This increases vulnerability and limits culturally sustainable climate adaptation and possibilities to react to exceptional situations, especially in situations where technology cannot help, is unavailable, or broken.

Cultural adaptation to the climate change

Adaptation measures started in the 1990s and 2000s (see Tables 6.1 and 6.2). In reindeer work model 4, in the boreal region, supplementary feeding of reindeer and a change in the reindeer work model began even before the effects of climate change became significant. This reindeer work model developed as a response to competitive land use, especially to forestry and loss of pastures. The change has created a basis for adaptation to climate change. Saami reindeer herding is very diverse and adapts to social and environmental changes through a reindeer work model, for example, introducing new innovations or changing the pasture cycle to meet new challenges.

> Adapting to climate change is about the future of Saami culture. The reindeer livelihood maintains the Saami identity, language, and culture. They are the roots that unite different generations and create cultural continuity.
>
> (Informant 9)

The physiology and the nature of the reindeer create restraints on adaptation. Certain tasks, such as separations and calf marking, must be done at a specific time, and the nutritional needs of reindeer also set boundaries on adaptation.

Drones are considered as means for adaptation to climate-induced forestification and shrubification. The use of drones is at an experimental stage, used mainly to track and sometimes even herd reindeer in reindeer work model 6. Drones are considered practical when reindeers are in shrugs or dense forest. Drone users have stressed that the key benefit of using them is that drones do not leave traces on the terrain, unlike ATVs. The use of drones not only requires technical know-how but also knowledge of reindeer behavior and landscape, so that drones herd reindeer into the desired area.

> We Saami have been pioneers in adaptation. Reindeer herding is a really good thing. Scientists say that when reindeer graze, the world does not warm up so much, that it is important. And second, young herders have begun to incorporate new technologies. They use both drones and dogs to herd the reindeer. These things do not erode nature nor pollute. Such small actions can be used to adapt.
>
> (Informant 9)

The interviewees are highly aware of climate change and environmental effects on their livelihood. They, like the researchers, have deduced that reindeer herding is important for climate mitigation. Reindeer grazing can regionally counter-impact the climate-induced shrubification in the arctic and sub-arctic region (Kaarlejärvi et al. 2015), and reindeer grazing in summer can be used to delay snowmelt, increase surface albedo, and decrease the ground heating in the snowmelt season (Cohen et al. 2013).

The limits of adaptation have also been considered, in other words, how many of the traditions may be lost. There is a strong belief in the future of reindeer herding as a business, but there is great concern about what will be lost in the process of adaptation, and whether the Saami way of life and culture will survive.

> We have been adapting for a long time. Soon, it will not be Saami reindeer herding anymore if the changes continue. That change is a pretty radical thing that is going on all the time.
>
> (Informant 10)

Our interviews show that the individual herders are left alone to secure their livelihoods in a changing climate, without any support from larger society. They have experienced that they, and their way of life, are guided by social obligations, however, without the authorities understanding the impact of their decisions on reindeer culture in the long run. Informants were concerned about the legal and administrative harmonization of reindeer herding; ways should be found to prevent and slow this development. If the differences in reindeer work models and their specific needs for adaptation are not understood and recognized, there is a real danger that climate policies and measures planned by authorities could pursue and support a standardized reindeer work model.

> We have been left alone to cope with climate change. They (authorities and researchers) want us to reduce the number of reindeer and maintain reindeer herding without supplementary feeding, or to increase supplementary feeding or graze reindeer on natural pastures without eroding vegetation. No matter what you do, you always do it wrong. Then they want our reindeer to be prey for the predators. There's always discussion on limiting our grazing rights and number of reindeer. At the same time, they want new areas for tourism, railroads, forestry, or something else. We can only protest and try to adapt. It is easy to blame us for everything, but no one takes responsibility for helping us with climate change adaptation. I fear for the future.
>
> (Informant 12)

Overall, study interviewees report that attitudes, governance, and legislation are barriers to adaptation. They find that the government does not take climate change and altered conditions into consideration, nor Saami reindeer herding culture, thereby posing barriers for cultural adaptation (see Eira et al. 2018; Furberg et al. 2011). For example, informants felt that the herders' experience, skills, and culture are downplayed compared to scientific knowledge. Saami and their needs are contrasted with the needs of all of Finland, the economy, and competing land uses. Informants are frustrated: they have a lot of knowledge on climate change; however, this knowledge is not appreciated by authorities. The interviews revealed that reindeer herders want to become more widely involved in climate policy, but that this work also needs more support.

Discussion and conclusion

Saami reindeer herders are facing a very complex set of challenges, posed by legislation and cultural change, due to their small population size, competing land uses, infrastructural development, public opinion, and pasture conditions. With increasing climatic changes, herders must balance all these external forces and pressures, in addition to climate change adaptation within their cultural traditions and values to secure economic and cultural survival. Competing land uses increase competition within a reindeer herding cooperative, to the point that reindeer work sometimes has negative effects on the sense of community within a cooperative.

Based on our findings, climate adaptation is a process of cultural change in response to climate-induced changes in the environment and society. Landscape memory (the cultural core of a shared knowledge system) and livelihood models are changing and evolving to adapt to the new situation. Saami climate cultures do not develop passively, as reindeer herders actively seek and choose the right adaptation methods for their community. Landscape memory is also a tool for perceiving and adapting to the changes in the environment, and for monitoring the effects of climate change. Saami knowledge about climate change and its impacts is produced and interpreted through landscape memory, from the viewpoint of their livelihood and traditions. Climate cultures among the Saami are highly specialized and contextual. Landscape memory theory can help to understand and analyze how specific climate cultures develop in this overall context.

Reindeer herding as a livelihood has proven its ability to adapt to climate change, but adaptation measures require resources that cannot be met by reindeer meat sales alone. Resources must be made available through culturally acceptable subsidies, and through a diversification of livelihoods. If state control and administration over reindeer herding increases, the cultural possibilities for reindeer herders to adapt to climate change may be weakened and reindeer herding models would likely become uniform. Based on the results of this research, it is important to stress that reindeer herders are involved in climate change mitigation measures, reindeer herding subsidy systems, and management to enable cultural adaptation to climate change.

Adaptation is a cultural process that includes selection and rejection (Fiske et al. 2014, 41–48). When discussing the diversity of reindeer work models (Table 6.1), it can be observed that the *siidas* have pursued different ways of adapting to new situations. All *siidas* have similar sets of options to use supplementary feeding or technology in their reindeer work; however, various adaptation methods have been chosen or rejected, considering pasture conditions and landscape memory. Herders have adapted their culture and livelihood to altered conditions in their own landscape, selecting some and rejecting other options, resulting in different reindeer work models that continue developing and adapting to new challenges. We also found that possibilities for adaptation to climate change are limited at the level of a single herding cooperative. The extent of competing land uses limits the possibilities for climate change adaptation, especially in the boreal region.

This study has shown that, in order to understand cultural impacts of climate change and possibilities for adaptation, local perceptions and experiences are important and reliable. Cultural effects of climate change can be examined through systematic climate ethnography. Our study indicates that Saami indigenous climate cultures are subject to change due to rapid changes in the landscape, biodiversity, and weather conditions. The socially and culturally shared knowledge of climate change includes diverse understandings of causality and intergenerational effects. The shared knowledge of climate change observed in landscape memory – in Saami climate cultures – is common among Reindeer Saami communities that have developed, and are developing, different reindeer work models (Figure 6.2). Although different reindeer work models exist, Saami reindeer herders have observed similar effects of climate change on the environment and on weather, and have identified impacts on their culture and livelihoods, indicating that landscape memory is a culturally shared holistic knowledge system that plays a key role in climate perception and adaptation.

Our study shows that climate-induced changes in biodiversity and weather conditions have significant and far-reaching socio-cultural implications for the Reindeer Saami communities. Our application of climate ethnography (Crate 2011) adds an important aspect to the climate cultures debate in this volume: the question of how climate change and climate adaptation itself induces cultural change. For the Saami, a central aim of reindeer herding is to maintain and transmit the Saami culture to future generations, much as previous generations have done. A crucial question for the future of the Saami is how the communities can adapt to climate change in a culturally sustainable way, mitigate the risks and losses brought by climate change, and, ultimately, how society at large can support this adaptation. Climate change adaptation requires a balancing of cultural traditions and values, administration, and legislation, it has to weigh sufficient income and survival against increasing pressures and stress. The study shows that reindeer herders do not passively react to changes but try to actively adapt their reindeer work to altered circumstances and conditions; we can see that communities have made different kinds of choices to maintain their livelihood (compare with Pelling 2011, 34–36, 56–60).

We conclude that for Saami reindeer communities, climate change adaptation is a process of cultural change in response to changes in the environment and society. The adaptation process manifests itself in different reindeer work models. The greatest challenges facing reindeer herders are the uncertainty and unpredictability of the new conditions which may undermine their landscape memory and its significance. Landscape memory has been the central tool for reindeer herders to maintain their cultural livelihood and adapt to the changes in the surrounding environment. If landscape memory decays, future generations will be left with fewer possibilities for adaptation and a thinned-out knowledge base. This study has shown that reindeer herding as a *cultural form* is highly vulnerable to the effects of climate change. However, *reindeer herding as a livelihood or economy* is flexible and can adapt to adverse effects of climate change; there are possibilities for sustainable economic adaptation with the help of technology and supplementary feeding.

Ethics statement

The *SAAMI* study has received free, prior, and informed consent by the Saami Parliament of Finland and by Skolt Saami Village Assembly (Sámediggi 2016). Participants in the study have given their written consent to participate in the project and have been informed of the study results in North Saami and Finnish.

Acknowledgments

The study *SAAMI* (*Adaptation of Saami to Climate Change*) was supported by a Finnish Government Grant (VN TEAS VNK, 48/49/2019), by the Finnish Cultural Foundation (for the *Arahat Project*), and by a personal Research and Travel Grant for Klemetti Näkkäläjärvi from the Jenny and Antti Wihuri Foundation. The study was implemented in 2019–2020. Interviews in the *Kaldoaivi* and *Paistunturi* Reindeer Herding Cooperatives, and a portion of the interviews in the *Muotkatunturi* Cooperative, were conducted by Joni Saijets. All other interviews were conducted by Klemetti Näkkäläjärvi. The authors would like to thank the informants of the study for their valuable input and collaboration, as well as the Saami Parliament and Skolt Saami Village Assembly for their support. We would also like to thank the peer reviewers and editors for reviewing the chapter. We sincerely appreciate all their valuable comments and suggestions which helped us to improve the quality of the text. The editors would like to thank Hans-Joachim Gruda (Berlin/Jukkasjärvi) for his support and for his commitment to finding researchers related to Sápmi for this volume.

References

Archer, Lewis, James D. Ford, Tristan Pearce, Slawomir Kowal, William A. Gough, and Mishak Allurut. 2017. "Longitudinal Assessment of Climate Vulnerability: A Case Study from the Canadian Arctic". *Sustainability Science* 12 (1): 15–29.

Clifford, James 1986. "On Ethnographic Allegory". In *Writing Culture: The Poetics and Politics of Ethnography*, edited by James Clifford and George Marcus, 98–121. Berkeley: University of California Press.

Cohen, Juval, Jouni Pulliainen, Cécile B. Ménard, Bernt Johansen, Lauri Oksanen, Kari Luojus and Jaakko Ikonen. 2013. "Effect of Reindeer Grazing on Snowmelt, Albedo and Energy Balance Based on Satellite Data Analyses". *Remote Sensing of Environment* 135: 107–117.

Convention on Biological Diversity. 1992. *10 years on taking stock, looking forward.* https://www.cbd.int/doc/legal/cbd-en.pdf

Crate, Susan A. 2011. "Climate and Culture: Anthropology in the Era of Contemporary Climate Change". *Annual Review of Anthropology* 40 (1): 175–194.

Cunsolo Willox, Ashlee, Sherilee L. Harper, James D. Ford, Karen Landman, Karen Houle, and Victoria L. Edge. 2012. "'From This Place and of This Place'. Climate Change, Sense of Place, and Health in Nunatsiavut, Canada". *Social Science and Medicine* 75 (3): 538–547.

Eira, Inger Marie Gaup, Anders Oskal, Inger Hanssen-Bauer, and Svein D. Mathiesen. 2018. "Snow Cover and the Loss of Traditional Indigenous Knowledge". *Nature Climate Change* 8 (2): 928–931.

Fiske, Shirley. J., Susan A. Crate, Carole L. Crumley, Kathleen Galvin, Heather Lazrus et al. 2014. *Changing the Atmosphere. Anthropology and Climate Change. Final Report of the AAA Global Climate Change Task Force.* Arlington, VA: American Anthropological Association.

Furberg, Maria, Birgitta Evengård, and Maria Nilsson. 2011. "Facing the Limit of Resilience: Perceptions of Climate Change among Reindeer Herding Sami in Sweden". *Global Health Action* 4: 1–11.

Heimann, Thorsten. 2019. *Culture, Space and Climate Change. Vulnerability and Resilience in European Coastal Areas.* New York: Routledge.

Herskovits, Melville J. 1958. *Cultural Anthropology.* New York: Alfred A. Knopf.

Ingold, Tim. 2000. *The Perception of the Environment: Essays on Livelihood, Dwelling and Skill.* London (United Kingdom): Routledge.

IPCC. 2014. *Synthesis Report. Contribution of Working Groups I, II and III to the Fifth Assessment Report of the Intergovernmental Panel on Climate Change.* Geneva (Switzerland): IPCC.

Jaakkola, Jouni J. K., Suvi Juntunen, and Klemetti Näkkäläjärvi. 2018. "The Holistic Effects of Climate Change on the Culture, Well-being, and Health of the Saami, the Only Indigenous People in the European Union". *Current Environmental Health Reports* 5 (4): 1–17.

Kaarlejärvi, Elina, Katrine S. Hoset, and Johan Olofsson. 2015. "Mammalian Herbivores Confer Resilience of Arctic Shrub-Dominated Ecosystems to Changing Climate". *Global Change Biology* 21 (9): 3379–3388.

Käyhkö, Jukka, and Tim Hortskotte, eds. 2017. *Reindeer Husbandry under Global Change in the Tundra Region of Northern Fennoscandia.* Turku (Finland): University of Turku.

Keskitalo, E. Carina H. 2008. *Climate Change and Globalization in the Arctic: An Integrated Approach to Vulnerability Assessment.* London (United Kingdom): Earthscan.

Lavrillier, Alexandra, and Semen Gabyshev. 2018. "An Emic Science of Climate: A Reindeer Evenki Environmental Knowledge and the Notion of an Extreme Process". *EMSCAT* 49. http://journals.openedition.org/emscat/3280

Mikkonen, Santtu, Marko Laine, H. M. Mäkelä, Hilppa Gregow, Heikki Tuomenvirta, M. Lahtinen, and Ari Laaksonen. 2015. "Trends in the Average Temperature in Finland 1847–2013". *Stochastic Environmental Research and Risk Assessment* 29 (6): 1521–1529.

Moerlein, Katie J., and Courtney Carothers. 2012. "Total Environment of Change". *Ecology and Society* 17 (1): 10.

Näkkäläjärvi, Klemetti. 2002. "The Siida, or Sámi Village as the Basis of Community Life". In *Siiddastallan. From Lapp Communities to Modern Sámi Life,* edited by Jukka Pennanen, and Klemetti Näkkäläjärvi, 114–121. Inari (Finland): Sámi museum.

Näkkäläjärvi, Klemetti. 2013. *"Jauristunturin poropaimentolaisuus: kulttuurin kehitys ja tietojärjestelmä vuosina 1930–1995"* [Reindeer Nomadism of Jávrrešduottar: Cultural Development and Knowledge System in 1930–1995]. Oulu (Finland): University of Oulu.

Näkkäläjärvi, Klemetti, and Suvi Juntunen. 2022(forthcoming). "Co-production of Knowledge on Climate Change Adaptation in Reindeer Saami Culture – Research Methodology and Ethics" (Submitted manuscript).

Näkkäläjärvi, Klemetti, Suvi Juntunen, and Jouni J. K. Jaakkola. 2020. *"SAAMI – Saamelaisten sopeutuminen ilmastonmuutokseen – hankkeen tieteellinen loppuraportti"* [Final Scientific Report of the Project SAAMI – Adaptation of Saami People to the Climate Change]. Helsinki (Finland): Valtioneuvoston kanslia.

Pearce, Tristan, Barry Smit, Frank Duerden, James D. Ford, Annie Goose, and Fred Kataoyak. 2010. "Inuit Vulnerability and Adaptive Capacity to Climate Change in Ulukhaktok, Northwest Territories, Canada". *Polar Record* 46 (2): 157–177.

Pekkarinen, Antti-Juhani, Jouko Kumpula, and Olli I. Tahvonen. 2015. "Reindeer Management and Winter Pastures in the Presence of Supplementary Feeding and Government Subsidies". *Ecological Modelling* 312: 256–271.

Pelling, Mark. 2011. *Adaptation to Climate Change: From Resilience to Transformation*. New York: Routledge.

Prno, Jason, Ben Bradshaw, Johanna Wandel, Tristan Pearce, Barry Smit, and Laura Tozer. 2011. "Community Vulnerability to Climate Change in the Context of other Exposure-Sensitivities in Kugluktuk, Nunavut". *Polar Research* 30: 7363.

Psathas, George. 1972. "Ethnoscience and Ethnomethodology". In *Culture and Cognition: Rules, Maps, and Plans*, edited by James P. Spradley, 206–222. New York: Chandler.

Rees, W. Gareth, Florian Stammler, Fiona S. Danks, and Piers Vitebsky. 2008. "Vulnerability of European Reindeer Husbandry to Global Change". *Climatic Change* 87 (1): 199–217.

Reindeer Herder's Association. 2019. *"Saamelaisten kotiseutualueen porotilastot porovuoteen 2018/2019"* [Reindeer Statistics in the Saami Home Region until Reindeer Year 2018/2019]. Paliskuntain Yhdistys, Rovaniemi (unpublished statistics).

Reindeer Husbandry Act. 1990. *Poronhoitolaki 848/1990*. https://finlex.fi/fi/laki/ajantasa/1990/19900848

Riedlinger, Dyanna, and Fikret Berkes. 2001. "Contributions of Traditional Knowledge to Understanding Climate Change in the Canadian Arctic". *Polar Record* 37 (203): 315–328.

Ruosteenoja, Kimmo. 2016. "Climate Projections for Finland under the RCP Forcing Scenarios". *Geophysica* 51 (1–2): 17–50.

Sámediggi. 2016. *Procedure for Seeking the Free, Prior, and Informed Consent of the Sámi from the Sámi Parliament in Finland for Research Projects Dealing with Sámi Cultural Heritage and Traditional Knowledge and other Activities That Have or May Have an Impact on This Heritage and Knowledge.* https://www.samediggi.fi/procedure-for-seeking-consent-for-research-projects/?lang=en

Savo, Valentina, Dana Lepofsky, J. P. Benner, Karen E. Kohfeld, J. Bailey, and Ken Lertzman. 2016. "Observations of Climate Change among Subsistence-Oriented Communities around the World". *Nature Climate Change* 6 (5): 462–473.

Part IV
North America

7 Contested Climate Cultures

Frame Resonance Disputes within the US Environmental Movement over Geoengineering Proposals

David Zeller

Introduction

Conceptions of vulnerability and resilience in the face of climate change are socially constructed (Christmann and Ibert 2012). How can frame resonance disputes illuminate the development and maintenance of socially shared knowledge constructions about climate change (Heimann 2018)? So-called "geoengineering" proposals occupy an increasing share of the global climate discourse, and controversies surrounding these potential solutions and adaptations offer the opportunity to examine contested climate cultures. Frames are discursive tools people use to make meaning from situations (Goffman 1974). This chapter argues that a more nuanced understanding of geoengineering framings within the US environmental movement – specifically, the frame disputes (Benford 1993) that emerged during a formative period in the development of the geoengineering discourse – is crucial for those who wish to understand recent efforts to shape the global politics of climate change. Analyzing how the US environmental movement constructs meaning around geoengineering proposals shows the difficult work involved in developing and maintaining socially shared knowledge constructions about climate change (Heimann 2018; Heimann and Mallick 2016; Christmann and Ibert 2012) – even among allies.

Four main findings resulted from the analysis. First, the analysis presented in the following shows how radical and reformist elements of the movement engaged in "boundary framing" practices (Phillips 2019; Hunt et al. 1994) in their online disputes about how the movement ought to portray geoengineering. Second, the analysis shows how activists adopt a strategy of simplicity in their attempts to render complex climate solutions understandable for their audiences. Third, the study shows how discursive flexibility may serve as a crucial resource during periods of uncertainty. Fourth, geoengineering proposals remind us that it is not possible to separate knowledge construction processes from spatial concerns.

Frame disputes within the US environmental movement are episodes that occur when conceptions of climate change vulnerability and resilience practices (Heimann and Mallick 2016) are not shared. Allied groups whose

DOI: 10.4324/9781003307006-12

conceptions of vulnerability and resilience are not shared because they are still in-the-making offer the opportunity to document and analyze the socially constructed and contested character of climate change knowledge (Christmann and Ibert 2012). Environmental movement discourse changed over time as activists grappled with the implications of geoengineering proposals. These changes in the discourse point toward the difficulties of maintaining a shared understanding about climate change, even among parties who are ostensibly "on the same team" in most respects. Analyzing changes in activists' perceptions about climate change vulnerability and resilience is important for our understanding of climate cultures because such research provides evidence of the contingency of climate change constructions.

There are two broad approaches to geoengineering that are commonly discussed. The first is known as solar radiation management (SRM), and involves blocking the absorption of solar radiation by spraying aerosols into the atmosphere or constructing large reflective surfaces in space, among other proposed methods. The second is known as carbon capture and storage (CCS) and describes a range of actions designed to remove excess carbon dioxide from the atmosphere and sequester it using various means. Both kinds of climate engineering proposals have generated strong scientific and public opinion. People discuss geoengineering (often also referred to as "climate engineering") in the comments section of online news articles devoted to the subject, and on the webpages and blogs of environmental movement organizations.

During the 2005–2015 time period chosen for this study, geoengineering became increasingly mentioned as a possible response to the issue of climate change. As my research shows, journal articles, reports authored by various scientific bodies, and news reports about geoengineering experiments tended to produce contentious framing activities as EMOs sought to develop and refine their framing of this issue over time amid an atmosphere of uncertainty. This chapter analyzes the framing activities of EMOs during the latter half of the study period as they make sense of controversial geoengineering proposals and attempt to articulate an understanding of geoengineering that will resonate with various audiences. EMOs frequently disagreed with other EMOs over what geoengineering means and how it should be discussed – or whether it should be discussed at all.

Crucially, my analysis suggests that paying particular attention to disputes over the manner in which organizations attempt to affect the interpretation of various audiences foregrounds highly interactive knowledge construction processes (Christmann and Ibert 2012) and avoids characterizing frames as uncontested, static, behind-the-scenes byproducts of discursively monolithic communities. Since frame resonance disputes are about the "how" of framing, analyzing them provides a window into the consequential work involved in creating and maintaining a cohesive climate culture. In other words, frame resonance disputes are empirical evidence of the contested character of climate cultures. Indeed, movement elites and others often disagreed publicly about the meaning of geoengineering proposals.

For the environmental movement, constructing shared geoengineering-related knowledge requires framing diagnoses and prognoses in a way that will strike a responsive chord. As will be shown in the following, discursive diversity within the US environmental movement serves as both a resource and a restriction in this regard. Understanding how radicals and reformists frame geoengineering is an important contribution to our understanding of these nascent proposals. Further, studying frame disputes within the environmental movement emphasizes the contested character of climate cultures by showing the ongoing negotiations that occur within these groups as they seek to make sense of controversial climate change adaptation measures. After a brief review of selected literature and a discussion of the methods used in the study, the results of the analysis are presented and discussed as they pertain to previous findings from the study of climate cultures.

Literature review on framing and frame disputes within the environmental movement

This section reviews selected literature on framing and frame disputes within the environmental movement. Frames are contested carriers of cultural meaning, and analyses that utilize the concept of the frame dispute (Benford 1993; Goffman 1974) are particularly well-suited toward contextualizing our understanding of the generation and negotiation of shared knowledge (Heimann 2018). Frames are used to align individuals with social movements and facilitate participation and mobilization for collective action (Snow et al. 1986). Framing involves the development and maintenance of diagnostic frames, prognostic frames, and frame resonance. Diagnostic frames "attribute blame for some problematic condition by identifying culpable agents, be they individuals or collective processes or structures", while prognostic frames provide "both a general line of action for ameliorating the problem and the assignment of responsibility for carrying out that action" (Snow and Benford 1992, 137). Frame resonance refers to the degree to which a social movement has the potential to persuade the uncommitted to commit to the cause. Since they are interpretive matters, all framings are subject to contention. In other words, disputes sometimes erupt while people are engaged in framing activities.

A few different types of frame disputes have been identified in previous research. In line with the framing tasks outlined in previous studies of frame alignment, Benford's (1993, 679) research on the nuclear disarmament movement differentiates between three kinds of intramovement frame dispute – "disagreements over what is [diagnostic], over what ought to be [prognostic], and over how to represent a movement's versions and visions of reality [frame resonance]". He also distinguishes *intra*organizational frame disputes from *inter*organizational frame disputes. Of the latter, he asserts that:

> coalitions of movement organizations are particularly conducive to frame disputes because they are comprised of activists from a variety of organizations, each having its own version of reality, agenda, and views regarding the ways in which the movement should go about the business of recruitment, activation, and contention for power.
>
> (Benford 1993, 680 f.)

This finding seems to be supported by much of the subsequent research on frame disputes within the environmental movement. Frame disputes often occur between radical and reformist groups within a movement (King 2008; Benford 1993), and intramovement factions sometimes engage in "boundary framing" activities (Hunt et al. 1994). These discursive demarcations reinforce distinct collective identities (Gamson 1997), enabling environmental activists to draw on familiar vocabularies that help them to make sense of geoengineering proposals. By reinforcing their collective identities through their framings of geoengineering, the factions that make up the movement seek to foreclose or foretell certain climate change futures.

For the environmental movement, disputes about the influence of carbon dioxide emissions on the atmosphere and whether or not human beings bear responsibility for climate change are relatively rare. While diagnostic frame disputes may occasionally occur during periods where information about an issue is scarce (for example, see Futrell 2003), prognostic frame disputes and frame resonance disputes seem to be common occurrences within the environmental movement.

For example, Wahlström et al. (2013) surveyed environmental protestors in Copenhagen, Brussels, and London in order to determine how and why the collective action frames employed by demonstrators varied. Descriptive analyses revealed that protestors in Brussels favored "changing individual attitudes and behavior" while the Copenhagen and London demonstrators advocated legislative and policy changes over individual behavioral changes. Global justice and system change framings were infrequently invoked by the individuals they surveyed. Wahlström et al. (2013, 119) juxtapose individual beliefs with organizational beliefs, reiterating the importance of ongoing frame alignment efforts and the difficulty inherent in bringing "everyone involved into the same frame". Similarly, research by Emilsson et al. (2020) found frame disputes between Swedish protestors who largely ranked environmental concerns ahead of economic growth, noting the complexity with which individuals' competing conceptions interact.

In contrast to Wahlström et al. (2013), Walker (2009, 356) argues that the environmental justice frame has become an influential and flexible frame for understanding socio-environmental issues, "subject to necessary but sometimes problematic processes of recontextualization".

The strength of the environmental justice frame, Walker argues, can be seen in its recent extensions. In South Africa, the diffusion and adoption of the environmental justice frame proceeded along rhetorical lines that are familiar to Americans:

> the connections between the civil rights movement in the USA and anti-apartheid struggles in South Africa meant that the discourse of environmental racism resonated strongly in a country where the racialization of space had been institutionally organized and maintained through state power.
>
> (Walker 2009, 367)

Walker (2009, 369) notes that "the environmental justice frame is not singular, but rather flexible and dynamic, open to reconstruction as it moves both in space and time". The discursive flexibility of frames may prove crucial in times of uncertainty, including during deliberations about potential climate solutions. During these unsettled situations, however, factions within a movement sometimes clash.

Frame disputes frequently seem to occur as a result of tensions between radical and reformist visions of change. King (2008) notes that factionalization resulting from frame disputes does not necessarily portend disaster for movement organizations, however. The Sierra Club factions that split over the issue of immigration exhibited striking similarities to the factions Benford examined in his study of the nuclear disarmament movement in that the dispute hinged, in part, on whether the groups ought to pursue a single-issue or multi-issue strategy (King 2008, 56).

King (2008) found that the unique structure of the Sierra Club allowed it to endure and thrive during this decades-long intramural conflict over immigration policy. Factions were able to work toward other organizational goals despite the dispute. In some cases, it appears that frame disputes, even those of extremely long duration, do not necessarily produce problems for movement organizations.

In sum, frame disputes lay bare the subtle contours of discourse within a movement. As a concept, then, the frame dispute creates conceptual space for researchers interested in detailing the influence of intramovement contention on the accomplishment of discursive solidarity. This is because the very act of identifying a frame dispute requires an appreciation for the difficulties inherent in frame maintenance, foregrounding the dynamic character of framing by emphasizing the interactive elements of their always-ongoing negotiation. The use of the term "the" in the widely used phrase "the US environmental movement" suggests that that a single climate culture exists among these environmentalists. But if a shared understanding of climate change vulnerability and resilience is a central component of climate cultures, it is clear that "the" US environmental movement actually contains various internal subdivisions. Thus, the disputes I analyze in the following occur between different climate subcultures, so to speak, which exist within the larger climate culture known as the US environmental movement.

Methods

EMOs often post blogs, newsletters, photos, videos, and other forms of movement literature on their websites that are relevant to their portrayals of geoengineering. My selection of EMOs for analysis was theoretically motivated and is not intended to be a representative sample of the environmental movement discourse surrounding geoengineering. Similarly, the methodological decisions taken in this study were not intended to provide insight into the behind-the-scenes strategic calculations that inform the positions of activists. Since all of the data in the present study were available publicly, none of the names mentioned in the analysis have been anonymized. The EMOs analyzed were chosen from two comprehensive surveys of environmental movement organizations in the United States (Brulle et al. 2007, 89; Bosso and Guber 2005). From these studies, I selected EMOs that were previously familiar to me either from media coverage or from the scholarly literature on the environmental movement. Any EMO that produced English-language online content about geoengineering during the 2005–2015 period was eligible for inclusion in the study. Organizations that did not produce online content about geoengineering were excluded. Using a popular Internet search engine to reach the webpages of these EMOs, I then proceeded to search within the organizations' websites using key geoengineering-related terms and phrases. This ultimately resulted in a total of 16 EMOs who had discussed geoengineering online during the study period. In all, this data collection strategy resulted in an original dataset consisting of 86 distinct webpages or other online documents that mention geoengineering proposals. Table 7.1 lists the EMOs in the study.

The 2005–2015 study period was chosen for a couple of reasons. First, geoengineering proposals had only recently entered the online environmental movement discourse in earnest during the early part of the period. Environmental researcher Paul Crutzen (2006) and environmental advocates such as Stewart Brand (2009) were beginning to explore these proposals as technologies of last resort and exhorted environmentalists to explore their potential in the face of an increasingly carbonized planet. Second, and relatedly, the decade-long study period was chosen to capitalize on the initial uncertainty that often results from new technological proposals and to maximize the possibility that variation in geoengineering-related framings might be observed.

I engaged in a careful and intensive frame analysis (Goffman 1974), examining data from all 16 EMOs for themes, focusing, in particular, on discourse concerned with representational strategy. The initial period of data analysis primarily consisted of an investigation of the discursive field surrounding climate engineering in order to establish generic categories of representation. Following Benford (1993, 682), frame disputes were coded

Table 7.1 EMO geoengineering discourse online, 2005–2015

Organization	Blogs	Press releases	Newsletters	Other	**Total**
350.org	2	1	0	0	**3**
Climate Reality Project	2	0	0	1	**3**
Earthjustice	1	0	0	1	**2**
Earth First	0	0	5	0	**5**
Environmental Defense Fund [EDF]	13	0	0	4	**17**
Friends of the Earth [FoE]	5	2	0	1	**8**
Greenpeace	6	0	0	1	**7**
National Audubon Society	0	0	2	0	**2**
National Wildlife Federation	1	0	0	0	**1**
Natural Resources Defense Council [NRDC]	0	1	0	2	**3**
Nature Conservancy	1	0	1	1	**3**
Rainforest Action Network	1	0	0	0	**1**
Sierra Club	5	0	5	8	**18**
Union of Concerned Scientists [UCS]	7	0	0	2	**9**
Wilderness Society	0	0	0	1	**1**
World Wildlife Fund	1	1	1	0	**3**
Total	**45**	**5**	**14**	**22**	**86**

Source: Own representation.
Note: "Other" types of documents included announcements, articles, book and film reviews, an eco-vocabulary quiz, interviews with authors, a "message from the chair", news briefs, a petition, Q&As, a report, a strategic plan update, and web pages.

according to three generic forms of framing activity: *diagnosis, prognosis,* and *frame resonance.* This often resulted in overlapping codes, for example, even relatively brief framings sometimes included both an evaluation of geoengineering as a solution (prognosis) and a discussion of how geoengineering should be represented (frame resonance). The segments of text coded as frame disputes were sometimes as short as a phrase, while others were a few paragraphs in length. More important than constraining the length of the segment was a desire to faithfully present the entire context of the disagreement. Thus, some coded segments were quite brief while others were extensive by comparison. This strategy made it possible to characterize the general state of the geoengineering discourse within the movement. In the remainder of the chapter, I focus on the results of my analysis of frame resonance disputes within the US environmental movement over geoengineering proposals, which tended to center on the "how" (i.e. manner) and the "who" (i.e. audience) of the framing activity.

Analysis

This section documents frame resonance disputes within the US environmental movement over geoengineering proposals. Controversies over climate change solutions such as geoengineering show that climate cultures require discursive maintenance. Framing requires work. Like other social movements, the US environmental movement consists of coalitions of similarly oriented organizations that nevertheless have varying aims and strategies for achieving change. Many EMOs produced discourse online about geoengineering proposals during the 2005–2015 study period, and their disagreements as to how these proposals ought to be discussed were, unsurprisingly, linked to their views about the viability and ethics of these proposals. In other words, their prognostic frame disputes also often included disputes over frame resonance. Thus, my analysis shows that their evaluations of these proposals as solutions influence the way that they feel about how they ought to be represented.

From 2005 to 2009, the US environmental movement had largely been unified in their condemnation of SRM geoengineering proposals as "dangerous distractions" from the root issue of climate change, namely, runaway carbon dioxide emissions. EMOs such as Friends of the Earth (FoE), the Nature Conservancy, the Sierra Club, and 350.org continued to regard geoengineering proposals this way through the end of the study period. Over time, other EMOs like the Environmental Defense Fund (EDF) and the Natural Resources Defense Council (NRDC) began to regard engagement with these proposals as a regrettable responsibility. In their view, lack of meaningful progress with regard to climate change required an understanding of various emergency climate intervention measures. For these EMOs, geoengineering proposals are unfortunate necessity brought about by the poor climate change decisions of the past.

Four main findings emerged from the analysis. First, EMOs engaged in boundary framing activities in their disputes over how best to represent geoengineering proposals. Second, EMOs adopted a strategy of simplicity on occasion, seeking to render these sometimes-complex proposals meaningful for their audiences. Third, the diversity of organizations that comprise the US environmental movement imbue the movement with a measure of discursive flexibility that may serve as a useful resource during uncertain times. Finally, geoengineering proposals remind us that it is impossible to decouple climate knowledge construction from spatial concerns; indeed, climate policy decisions often have effects far beyond the borders of those who implement them. While this aspect of geoengineering proposals was only rarely addressed by the EMOs in this study, I return briefly to the implications of this insight in the following section.

During the study period from 2010 to 2015, which was characterized by an expansion of definitions and a curtailing of discussions, frame resonance disputes were evident. In their disagreements as to the manner in which the movement should attempt to affect the interpretation of various audiences, the US environmental movement engaged in boundary framing practices. By this time, EMOs that framed geoengineering as a regrettable responsibility were advocating for the advancement of the geoengineering discussion, seeking to expand the boundaries of conventional definitions of geoengineering in the process. Meanwhile, EMOs who regarded these proposals as dangerous distractions often recommended restricting the boundaries of the conversation by not discussing these proposals at all. Their framings are alternately attempts to challenge or maintain fidelity with conventional understandings about climate change solutions within the movement. In this way, frame resonance disputes over geoengineering offer empirical evidence of the contested character of climate cultures.

Alongside, and despite pleas to exclude geoengineering from the climate change discourse, there also came calls to expand the definition of geoengineering. By December 2012, NRDC was among the EMOs who sought to expand the definition of geoengineering to include the unintended consequences of human behavior (e.g. carbon emissions) alongside intentional geoengineering methods. The Natural Resources Defense Council posted a contribution from science fiction author Kim Stanley Robinson that illustrates this framing. Note the way he anticipates the arguments of other well-meaning environmentalists, reframing geoengineering as a responsibility rather than a risk:

> Many people have expressed doubt that the proposals would work, or believe that a string of negative unintended consequences could follow. Merely discussing these ideas, it has been said, risks giving us the false hope of a 'silver bullet' solution to climate change in the near future – thus reducing the pressure to stem carbon emissions here and now. These are valid concerns, but the fact remains: our current technologies are already geoengineering the planet – albeit accidentally and negatively.

This framing aligns closely with the notion that "*the impossible is already happening*" as one commenter had put it five years earlier. By expanding the definition of geoengineering, he seeks to reframe hubris as hope. Robinson goes on to mention several other examples where environmental actions might "scale up" to the point where they "could be thought of as geoengineering". Among other examples, he mentions the pursuit of population stabilization by "promoting women's legal and social rights" since:

> wherever they expand, population growth shifts toward the replacement rate. This particular geoengineering technology nicely illustrates how the word technology can't be defined simply as machinery; it includes things like software, organizational systems, laws, writing, and even public policy.

Here, Robinson is not only trying to expand definitions of geoengineering but also of technology itself. It may be the author's status as a science fiction writer that affords him the ability to comfortably envision beneficial climate interventions on a planetary scale. Regardless, some commenters were un-impressed with the way that Robinson contested these normative definitions of geoengineering and technology.

When an EMO engages in reframing activities, these efforts take place in the context of already-existing framings. That means that when framings of geoengineering make use of new rhetoric that departs from the original script, resistance is likely to result. Interactions like those between Robinson and his readers are indicative of this type of resistance. For the vast majority of its existence, the environmental movement has largely remained steadfast in its framing of carbon dioxide emissions – the solution to the climate change problem requires the elimination of this "root cause". Some environmentalists are uncomfortable with the new narrative espoused by EDF, NRDC, and their discursive allies. They view geoengineering as, at best, an excuse to retain our current system of fossil fuel production and consumption.

However, while EMOs like EDF and NRDC turned toward a different understanding of geoengineering during the latter half of the period of analysis, other groups like Earth First, FoE, and 350.org continued to characterize geoengineering as a dangerous distraction.

For example, in early 2013, Earth First author "Ned Ludd" wrote a contribution to the organization's newswire to commemorate "the 200th anniversary of the Luddite uprisings". Geoengineering proposals, according to Ned, can be attributed to the "insane" hubris of capitalist technocrats. Instead of a technical assessment of geoengineering proposals, he argues, readers ought to "stand back for a moment, and realize where capitalist technocracy has brought us". He goes on:

> As industrial capitalism has taken us to the point of destabilizing the entire climatic system of the planet, scientists are now seriously proposing to intervene in that system, about which next to nothing is understood. The over-confidence and hubris of technocrats has led them to a point where they feel able to literally play God with the basic life support systems of the entire planet, despite the colossal but incalculable risks that such an enterprise would entail, for the planet and for those other humans, in countries which will have no say in whether these technologies are used.

Importantly for the present analysis, Ned mentions more than once that the best way to mobilize action around "techno-fixes" like geoengineering is by not discussing these proposals at all: "it's vital that the ecology movement does not even discuss such techno-fixes. We must simply say, 'No!' and be prepared to back that up with action". This tactic of non-discussion was used by EMOs like Earth First and 350.org in the final two years of the study period. These EMOs feel that the optimal manner in which the movement ought to represent geoengineering proposals is by representing them as unworthy of consideration – in other words, by choosing not to represent them at all.

This tension continued for the remainder of the study period. If geoengineering is a "dangerous distraction" subject to "unintended consequences", then it should not be considered as a potential solution. Some EMOs even went so far as to say that geoengineering should not be discussed at all. However, other activists argue that if geoengineering is effectively "already happening" because of the negative impacts of society on the environment, then perhaps society can harness the knowledge of our collective environmental impact toward a positive environmental outcome. Such a global project would necessarily entail careful and considerate deliberation.

The final year of the analyzed time period yielded the highest number of frame resonance disputes. The late onset of these kinds of disputes may have been partly an artifact of the processual nature of the core framing tasks as well as the inchoate state of geoengineering knowledge within the environmental movement. Indeed, it is impossible to observe any kind of frame dispute until an initial framing has been established and a subsequent framing differs from that initial framing. By the end of the analyzed period, however, discursive divisions within the movement over geoengineering were plainly evident.

For example, the National Research Council's newly released report on SRM geoengineering initiated a brief exchange between reader Ted Parson and Union of Concerned Scientists (UCS) blog author Peter Frumhoff early in 2015. After reading Frumhoff's take on the implications of the report, Ted offered a few suggestions with regard to frame resonance. One of the primary considerations for Ted was to first establish precise terminology before moving the geoengineering conversation forward. Ted notes that

> there's a lot of useless argument over what to call this, what to call the two major approaches within it – actively messing with the global carbon cycle, or actively messing with the Earth's radiation balance. For here, I'll call them 'carbon stuff' and 'sunlight stuff'.

Ted opts for simplicity ("carbon stuff and sunlight stuff") rather than a more elaborate description ("carbon recovery and albedo restoration") as he attempts to work out the optimal manner in which geoengineering proposals ought to be framed.

For their part, 350.org seems to side with EMOs like the Sierra Club, Greenpeace, FoE, Earth First, and other groups who generally oppose geoengineering. Though they did not produce a large amount of discourse about geoengineering during the period of study, a 350.org press release stated that

> (clean energy) solutions should not include geoengineering, nuclear, and other 'false-solutions' that impact poor and vulnerable communities.

With clean, just, and renewable energy sources so readily available, getting lost in a discussion about last ditch technologies to save the planet is a dangerous distraction.

To a certain extent, the frame resonance disputes I observed seemed to be the inevitable outgrowth of prognostic frame disputes. EMOs framing geoengineering in terms of risk frequently dismissed such proposals as "dangerous distractions" rife with the potential for "unintended consequences" EMOs framing geoengineering as a regrettable responsibility generally provided dry technical summaries of the various proposals, arguing that our negative impact on the planet should give us hope that we might be able to reverse course. As one NRDC blogger wrote in early 2015:

> for now, think of solar radiation management as the panic room of climate change mitigation. Few people really want to go there, but it's still tempting to some especially since our current rate of progress on carbon reduction suggests things could get pretty bad. Let's hope we can turn things around before it comes to that.

In sum, EMOs that regarded geoengineering as a dangerous distraction tended to shun discussions of these proposals, while EMOs that felt geoengineering was a regrettable responsibility advocated for additional deliberation.

Discussion and conclusion

This study yielded four major findings, at least two of which have direct implications for the study of climate cultures. First, so-called boundary framing practices influence internal climate knowledge construction processes within the US environmental movement (Phillips 2019; Hunt et al. 1994). Second, EMOs discussed adopting a strategy of simplicity in order to reduce the complexity of geoengineering proposals for activists; this finding converges with previous climate cultures studies (Heimann 2018; Heimann and Mallick 2016). Third, discursive flexibility may be a crucial resource during periods of nascent climate reality construction (Walker 2009). Fourth, discursive deliberations over geoengineering proposals serve as a reminder that it is impossible to decouple knowledge-related processes from spatial concerns (Heimann and Mallick 2016). I elaborate on these insights in this final section.

Frames define the terms of discourse, yet they are contested carriers of cultural meaning. EMOs attempted to draw discursive boundaries in their disputes about how geoengineering proposals ought to be represented. These frame resonance disputes offer further empirical evidence of the contested character of climate cultures. Paying attention to disputes over the manner in which organizations attempt to affect the interpretation of various audiences foregrounds highly interactive climate knowledge construction processes (Christmann and Ibert 2012), and avoids characterizing geoengineering framings as uncontested, static, behind-the-scenes byproducts of discursively monolithic communities. Further, frame resonance disputes over geoengineering largely revolve around challenges to conventional understandings of climate change within the movement. This also may have relevance for comprehensive understandings of the social construction of shared climate knowledge (Heimann 2018; Heimann and Mallick 2016; Christmann and Ibert 2012). Frame disputes over geoengineering proposals show that certain aspects of environmentalists' understandings of climate change are not shared. While this may pose a challenge for the US environmental movement, the socially constructed and contested character of climate change knowledge also reminds us that alternative climate futures are possible.

Frame disputes are not merely contests of semantic superiority. Rather, they are early opportunities to shape the still-forming conversation. As the boundaries of an issue become finite, the groups who make claims about the issue become distinguishable through their discursive productions. To the degree that disagreements over discursive boundaries disrupt established convention, those who would maintain such boundaries are compelled to hold the line. These processes described could aptly be characterized as "boundary framing" practices (Hunt et al. 1994, 194). To the extent that EDF, NRDC, and other EMOs seem to have embraced geoengineering, however reluctantly, this embrace is a noticeable departure from the traditional narrative about carbon dioxide espoused by the environmental movement. By refusing to discuss "techno-fixes", EMOs like Earth First and 350.org draw a discursive line that they argue ought not to be crossed. Activists should avoid "getting lost in discussions about last ditch technologies to save the planet", as a 350.org press release put it, because engaging in discourse about geoengineering may lend unwarranted legitimacy to the conversation. Geoengineering proposals are outside the boundary, restricted from the discussion. Boundary framing need not always result in a restriction of the terms of discussion, however. Sometimes boundaries are expanded as a result of these framing practices (Phillips 2019).

Frame disputes over geoengineering proposals complicate activists' understandings of sustainability, stewardship, vulnerability, resilience, and other issues central to global contemporary environmentalism. EMOs offered fragmented interpretations of socio-environmental reality via their geoengineering framings, and opted for simplicity in order to reduce complexity (Heimann and Mallick 2016) for activists, with varying success. Indeed, there were a few instances where EMO framings of geoengineering were challenged by one or more readers in the comments section below a blog post. Activists discussed reducing the complexity of the issue by substituting words like "stuff" for words like "restoration" and "enhancement". Readers decried the overly simplistic notion that the planet might be "engineered" at all.

The reticulated discourse surrounding geoengineering reminds us that the environmental movement, like all social movements, is comprised of multiple and sometimes-conflicting collective identities (Gamson 1997, 181). Some organizations within the movement are more accommodating toward technological solutions than others. The collective identity of these groups emphasizes a pragmatic approach to environmental issues. Other environmental organizations abhor "techno-fixes", and yet, both find common cause within a single, identifiable social movement.

Fortunately for the movement, discursive flexibility (Walker 2009) likely bolsters actual and potential membership, providing the space that is necessary for people to choose alternative paths toward the same goal. Indeed, when it comes to geoengineering proposals, it may be more useful to regard the environmental movement as a collection of disparate collective identities rather than as possessing a dominant, unified frame that the entire movement can be expected to coalesce around. This kind of discursive flexibility holds a great deal of utility for social movements and may help to explain the durability of the environmental movement. In fact, this type of flexibility may be necessary during contentious episodes of nascent reality construction.

This study documents the ways that climate knowledge about geoengineering is generated and negotiated. Frame resonance disputes provide insight into the discursive competitions that occur as climate cultures develop – contests that help determine the message sent by environmentalists to the public at large as well as other audiences. Frame disputes over geoengineering show the remarkable discursive breadth of the US environmental movement – even fundamental prognoses about carbon dioxide emissions are subject to dispute among allies, and these may point toward different strategies for achieving resonance. In short, frame disputes over climate change adaptations such as geoengineering consist of efforts to construct or resist the establishment of a new discursive order. Frame resonance disputes sometimes hinged on whether geoengineering proposals should be discussed at all. This strategy begs the question of whether predictions about future climate scenarios foreclose certain opportunities and nourish others. Put another way, can environmental utterances produce environmental realities?

Climate cultures are social constructions (Christmann and Ibert 2012), and like all social constructions, they are subject to negotiation and contention. In fact, it is their very fragility that renders them worthy of study. Concepts such as frame disputes can help to further contextualize and refine understandings of shared knowledge about climate change – including the difficult work involved in developing and maintaining shared meanings around potential solutions to climate change (Heimann 2018). Frame resonance disputes also show the dynamic aspects of framing activity. Rather than presenting frames as ready-made products, this study shows how they are continually re-made through interaction. Frame resonance disputes provide empirical evidence that climate cultures require ongoing maintenance.

As geoengineering research and governance initiatives continue in the coming years, the balance of the public conversation has shifted from questions of plausibility toward concerns about ethics and unintended consequences. The discourse reviewed in the present study largely exhibited a lack of concern about the potential for geoengineering schemes to produce environmental injustices. Though these concerns received brief mention in certain quarters of the movement, future deliberations about geoengineering ought to take these issues more seriously, as the impacts of these decisions will undoubtedly migrate across national borders and disproportionately affect already-disadvantaged populations. This has direct relevance for our understanding of climate cultures as well, as geoengineering proposals prevent the decoupling of knowledge-related processes from spatial concerns (Heimann and Mallick 2016). Indeed, geoengineering-related decisions made by nations or individuals may be global in their effects.

References

Benford, Robert D. 1993. "Frame Disputes within the Nuclear Disarmament Movement". *Social Forces* 71 (3): 677–701.

Bosso, Christopher J., and Deborah L. Guber. 2005. "Maintaining Presence: Environmental Advocacy and the Permanent Campaign". In *Environmental Policy: New Directions for the Twenty-First Century*, edited by Norman J. Vig, and Michael E. Kraft, 78–99. Washington, DC: CQ Press.

Brand, Stewart. 2009. *Whole Earth Discipline: An Ecopragmatist Manifesto*. New York: Viking Penguin.

Brulle, Robert, Liesel H. Turner, Jason Carmichael, and J. Craig Jenkins. 2007. "Measuring Social Movement Organization Populations: A Comprehensive Census of US Environmental Movement Organizations". *Mobilization* 12 (3): 195–211.

Christmann, Gabriela, and Oliver Ibert. 2012. "Vulnerability and Resilience in a Socio-Spatial Perspective: A Social-Scientific Approach". *Raumforschung und Raumordnung* 70 (4): 259–272.

Crutzen, Paul J. 2006. "Albedo Enhancement by Stratospheric Sulfur Injections: A Contribution to Resolve a Policy Dilemma?" *Climatic Change* 77 (3–4): 211–220.

Emilsson, Kajsa, Hakan Johansson, and Magnus Wennerhag. 2020. "Frame Disputes or Frame Consensus? 'Environment' or 'Welfare' First amongst Climate Strike Protestors". *Sustainability* 12 (3): 882.

Futrell, Robert. 2003. "Framing Processes, Cognitive Liberation, and NIMBY Protest in the US Chemical-Weapons Disposal Conflict". *Sociological Inquiry* 73 (3): 359–386.

Gamson, Joshua. 1997. "Messages of Exclusion: Gender, Movements, and Symbolic Boundaries". *Gender & Society* 11 (2): 178–199.

Goffman, Erving. 1974. *Frame Analysis: An Essay on the Organization of Experience*. Boston, MA: Northeastern University Press.

Heimann, Thorsten. 2018. *Culture, Space and Climate Change. Vulnerability and Resilience in European Coastal Areas*. London (United Kingdom): Routledge.

Heimann, Thorsten, and Bishawjit Mallick. 2016. "Understanding Climate Adaptation Cultures in Global Context: Proposal for an Explanatory Framework". *Climate* 4 (4): 112.

Hunt, Scott A., Robert D. Benford, and David A. Snow. 1994. "Identity Fields: Framing Processes and the Social Construction of Movement Identities". In *New Social Movements: From Ideology to Identity*, edited by E. Laraña, H. Johnston, and J. R. Gusfield, 185–208. Philadelphia, PA: Temple University Press.

Kahan, Dan M., Hank C. Jenkins-Smith, Tor Tarantola, Carol L. Silva, and Donald Braman. 2015. "Geoengineering and Climate Change Polarization: Testing a Two-Channel Model of Science Communication". *Annals of the American Academy of Political and Social Science* 658 (1): 192–222.

King, Leslie. 2008. "Ideology, Strategy and Conflict in a Social Movement Organization: The Sierra Club Immigration Wars". *Mobilization* 13 (1): 45–61.

Phillips, Ryan J. 2019. "Frames as Boundaries: Rhetorical Framing Analysis and the Confines of Public Discourse in Online News Coverage of Vegan Parenting". *Journal of Communication Inquiry* 43 (2): 152–170.

Snow, David A., E. Burke Rochford, Steven K. Worden, and Robert D. Benford. 1986. "Frame Alignment Processes, Micromobilization, and Movement Participation". *American Sociological Review* 51 (4): 464.

Snow, David A., and Robert D. Benford. 1992. "Master Frames and Cycles of Protest". In *Frontiers in Social Movement Theory*, edited by Aldon D. Morris, and Carol M. Mueller, 133–155. New Haven, CT: Yale University Press.

Wahlström, Mattias, Magnus Wennerhag, and Christopher Rootes. 2013. "Framing 'the Climate Issue': Patterns of Participation and Prognostic Frames among Climate Summit Protesters". *Global Environmental Politics* 13 (4): 101–122.

Walker, Gordon. 2009. "Globalizing Environmental Justice: The Geography and Politics of Frame Contextualization and Evolution". *Global Social Policy* 9 (3): 355–382.

8 The Politics of a Sustainable Coast

Competing Adaptation Cultures in Southeastern Louisiana

Michael A. Haedicke

Introduction

Climate change is simultaneously a physical process and a sociocultural phenomenon "whose meanings are very much in negotiation among social groupings of many kinds" (Callison 2014, 12). Often fraught, these negotiations are shaped by the emerging physical consequences of climate change and by the values and interests of participating groups (Adger et al. 2013; O'Brien and Wolf 2010). As the chapters in this volume demonstrate, the international literature about climate cultures has made significant progress in understanding how climate change takes on various meanings in different cultural communities, as well as exploring how socially constructed knowledge about climate change intersects with place-based efforts to adapt to the disruptions that changes in the global climate are likely to bring (Heimann and Mallick 2016; Christmann and Ibert 2012). This chapter offers two contributions to this literature.

The first contribution is empirical. Drawing from a set of interviews that I have conducted with environmentalists, state planners, and business leaders in southeastern Louisiana, I sketch out competing visions of climate adaptation that exist in this region of the United States. My analysis emphasizes divergent interpretations of a state-sponsored adaptation effort known as "The Master Plan for a Sustainable Coast". Acknowledging the dangers of sea level rise and severe weather, this Master Plan identifies a suite of environmental restoration and flood protection projects that may help to fortify communities and infrastructure against climate disruption. Since it has captured the lion's share of adaptation funding in Louisiana, the Master Plan has become a target of controversy. It therefore provides a valuable opportunity to study rival interpretations of climate risk and resilience.

The second contribution is theoretical. I extend the climate cultures framework, which is anchored in the constructionist sociology of knowledge, by bringing it into dialogue with scholarship in the political economy of the environment. This latter approach investigates how environmental activities – including those related to climate adaptation – are enmeshed in ongoing struggles between groups with different interests, resources,

DOI: 10.4324/9781003307006-13

and degrees of political influence (Carmin et al. 2015; Sovacool et al. 2015; Rudel et al. 2011). While climate cultures research tends to highlight interpretations that are shared among members of social groups, the political economy approach directs attention to cultural divergences, intergroup conflict, and inequalities of power. In Louisiana, this approach illuminates how different climate adaptation cultures are linked to the larger political and economic projects of social groups, such as local business leaders and progressive activists and scholars.

The next section of this chapter lays out this theoretical argument, explaining points of connection between research about climate cultures and studies of the political economy of the environment, and emphasizing how insights from the latter can complement and extend the climate cultures approach. Next, I describe coastal Louisiana as a distinctive place, considering its exposure to climate-related risks and the origins of the state's Master Plan. This is followed by a brief description of my research interviews and then by a more extensive discussion of key findings. I conclude with a reflection on how my approach to analyzing climate cultures in Louisiana can assist in research about climate adaptation elsewhere.

Climate cultures and the political economy of the environment

Climate adaptation cultures "can be understood as differences in perceptions of vulnerability as well as in preferred practices for creating resilience" (Heimann and Mallick 2016, 59). The climate cultures literature documents the character of these differences across geographic places and social groups, highlighting the fact that people in different cities or regions, or in different institutional fields in the same region, may express different interpretations of the perils and possibilities of climate change irrespective of the objective physical character of the impacts that they face (Heimann 2019; Christmann and Ibert 2012). Theoretically, this literature conceptualizes culture as shared knowledge – that is, as socially constructed perceptions of reality that are expressed and reproduced through interactions, discourse, and institutions (Berger and Luckmann 1967). Heimann (2019) further proposes conceptualizing culture as a "relational space", a concept that highlights how individuals with similar perceptions of climate risk and resilience cluster into discursive constellations that are formed through ongoing acts of interpretation and evaluation, as well as the extent to which these constellations map onto social space, such as geographic region.

In the work of Heimann and others, climate cultures are understood to derive from background knowledge constructions, defined as shared and largely taken-for-granted assumptions about current and desired future conditions of the world (Heimann 2019; Adger et al. 2013; O'Brien and Wolf 2010). In many cases, these background constructions are place-specific, anchored in the unique histories, experiences, and "local knowledge" of geographically identifiable communities (Gieryn 2000). Indeed, the intersections of climate change and place identity are profound. Not only can locally specific meanings shape perceptions of climate-related vulnerabilities and adaptation priorities, but climate change also has the potential to call existing understandings of a place's defining features and its relationship to other locations into question (Adger et al. 2011). Christmann et al.'s (2014) provocative examination of climate adaptation cultures in the German cities of Lübeck and Rostock illustrates the different dimensions of this relationship. Through an exhaustive analysis of climate adaptation discourse in local newspapers, they demonstrate that residents of Lübeck have interpreted climate threats through the lens of the city's historical identity and are particularly concerned to protect the traditional buildings in the city's center, which are seen as emblematic of that history. In the industrial town of Rostock, by contrast, climate change is associated with a shift in discursive attention away from the area's traditional fisheries and toward its beaches, which are seen as the locus of a future tourist economy made possible by warming temperatures.

Like the climate cultures literature, research that approaches environmental issues through the lens of political economy takes place seriously. However, the political economic approach emphasizes the social and cultural fragmentation of places, rather than their coherence. Places are conceptualized as arenas in which groups with differing interests, values, and degrees of political influence "struggle for control over the institutions and organizations" that determine the management of environmental resources (Rudel et al. 2011, 222). The distinctive character of places – including diverging cultural understandings of risk and resilience – is deeply entwined in these contests (Gotham 2016a). For example, an examination of the cases of Lübeck and Rostock through the lens of political economy might lead one to investigate the role of urban property owners in promoting a discourse of historical preservation (potentially leading to adaptation measures that protect their assets), or to ask questions about the inability of fishers to mobilize local government against the threats posed to their industry by climate change. As Logan and Molotch (1987, 43) put it, "places are not simply affected by the institutional maneuvers surrounding them. Places are those machinations".

Political economic research that investigates the origin and character of conflicts related to climate adaptation has yielded several important insights. First, it has revealed that adaptation projects often bring disproportionate benefits to economically and politically advantaged groups, reflecting their ability to influence the adaptation agenda and divert resources to serve their interests (Carmin et al. 2015; Sovacool et al. 2015). The approaches to resilience that these groups promote often reflect background values and beliefs that are shared by advantaged members of society. In the United States, for instance, such elite adaptation cultures emphasize business development and the protection of private property (Tierney 2014). Second, political economic research has directed attention to how excluded groups may contest elite adaptation cultures in the name of justice and fairness, offering an expanded understanding of the range of value commitments that may contribute to divergences between understandings of climate risk and preferred pathways to resilience (Adger et al. 2006). These contests show that "in climate change adaptation, as in development more generally, culture and politics interact to determine who has a voice, whose values count, and what information is legitimate" (Adger et al. 2013, 114).

When put side by side, the climate cultures and political economy approaches reveal the potential for fruitful synthesis (Heimann and Mallick 2016). While climate cultures research illuminates how shared values and beliefs lead to different interpretations of risk and resilience, the political economy perspective draws attention to how these interpretive differences intersect with social and political inequalities and intergroup struggles. In the analysis that follows, I advance a combined approach by exploring how diverging climate adaptation cultures in Louisiana correspond to political economic divisions between groups. I also investigate how groups' stances related to the Master Plan form part of broader political "meaning projects" (Fligstein and McAdam 2012, 54) in which they engage. That is, I consider how their statements about climate vulnerability and resilience link up with their positions on political issues that are not specific to adaptation, such as the proper relationship between industry and government and the relative weight that should be placed on economic development or social equity.

The place of coastal Louisiana

Coastal Louisiana is a place of contradictions, where the conservative, pro-business attitudes of the southern United States collide with rich traditions of environmental and social justice activism, and where increasingly dire climate-related risks are balanced against the state's economic dependence on onshore and offshore oil and gas extraction (Hochschild 2016; Freudenburg and Gramling 2012). A low-lying region facing the Gulf of Mexico, coastal Louisiana is exposed to damaging storm surges that result from frequent hurricanes and tropical storms. Climate models predict that storm-related flood risks will become more severe in coming years as the result of changing weather patterns and sea level rise (US Global Change Research Program 2018). Additionally, the marshy wetlands along the coast are subsiding (decreasing in elevation) at a relatively rapid rate. Subsidence exacerbates the risks posed by storms and sea level rise and has caused significant areas of the coast to convert to open water. Between 1932 and 2016, the surface area of coastal Louisiana decreased by approximately 2,000 square miles, an amount of land nearly equivalent to the US state of Delaware (Couvillion et al. 2017). Moreover, projections of future land loss under various sea level rise scenarios range from an additional 1,207 to 4,123 square miles by 2067 (Coastal Protection and Restoration Authority 2017, 72). In the worst case, one journalist recently noted, the iconic city of New

Orleans "could be left on a razor-thin sliver of land extending into the open Gulf, battered by storms rolling over the watery graves of unprotected communities" (Marshall et al. 2014).

Particularly in the eastern portion of the coast, which is an unstable delta of the Mississippi River, subsidence has an unavoidable natural component. Soil compaction, the decomposition of organic material, and activity along deep subsurface faults all contribute to this process (González and Törnqvist 2006). Nevertheless, there is widespread consensus that human activities have accelerated subsidence and are responsible for a good portion of coastal land loss (Yuill et al. 2009). During the early twentieth century, coastal Louisiana became a key node in the US oil and gas industry. Companies dredged an extensive network of access canals to drilling sites, which disrupted surface water drainage patterns and harmed marsh vegetation (Houck 2015; Theriot 2014). As much new oil exploration moved offshore in the 1970s and thereafter, the industry also built pipelines through wetlands areas, with further impacts on hydrological processes. Additionally, flood protection levees erected along the Mississippi River and its distributaries in the nineteenth and twentieth centuries have prevented river-borne sediments from replenishing wetlands. Instead of being spread out by episodic river floods, these sediments are channeled into the Gulf of Mexico (Paola et al. 2011).

Louisiana's current climate adaptation efforts seek to stop, or even reverse, subsidence and land loss as a means to increase the coast's resilience to climate-related changes. The initial version of the "Master Plan for a Sustainable Coast" was produced in 2007, following statewide discussions about the region's future after the disastrous 2005 hurricane season in which Hurricane Katrina devastated New Orleans and Hurricane Rita struck the southwestern coast. A primary strategy envisioned in this version and elaborated in later iterations is the construction of engineered diversions along the Mississippi and Atchafalaya Rivers that will allow sediments to flow into adjacent wetlands. These structures will be accompanied by new levee and floodwall systems around population centers and nonstructural initiatives, such as elevating and floodproofing buildings in vulnerable areas. The overall goal is to create "multiple lines of defense" (Lopez 2009) against floods and storm surges, building a redundant system rather than one that is dependent on a single component. As of this writing, the State of Louisiana projects that if all of the projects identified in the Master Plan are implemented, they will conserve approximately 800 square miles of land and reduce flood damages by about $8.3 billion over a 50-year period, as compared with a future in which these actions are not taken (Coastal Protection and Restoration Authority 2017).

As is the case with other transformations of the southern Louisiana landscape, the Master Plan represents a political project as well as a scientific and engineering one. It is supported by a coalition of moderate environmental groups, business leaders, and state actors. Capitalizing on the national attention paid to Louisiana in the wake of Hurricanes Katrina and Rita, these advocates were able to advance a series of state and federal laws that created a financial foundation for the Master Plan, as well as creating a state agency known as the Coastal Protection and Restoration Authority (CPRA), which is responsible for the Plan's implementation (Haedicke 2017). Under the 2006 Gulf of Mexico Energy Security Act (GOMESA), a federal law, Louisiana is allocated a portion of annual revenues received by the US government from the lease of oil exploration sites in the Gulf of Mexico. According to the state's constitution, these funds must be used to support projects in the Master Plan. Louisiana has also secured nearly $8 billion of the criminal and civil penalties paid by the energy company BP as a result of the explosion of the Deepwater Horizon oil drilling platform in 2010. Again, the bulk of this money is earmarked for the Master Plan.

However, some progressive environmental groups, university scientists, and coastal community representatives have been more critical of the Master Plan (Gotham 2016b). These critics generally acknowledge the need for climate adaptation measures but disagree with the approach that the state has embraced. These lines of political division thus represent boundaries between the different understandings of risk and resilience that make up diverging adaptation cultures in coastal Louisiana. Although disagreements sometimes appear to focus on the specific projects identified by the Master Plan, they also draw in different evaluations of the political processes and institutional arrangements that have informed the Plan's creation and diverging understandings of principles such as "fairness" and "sustainability".

Research approach

I investigated these divergent adaptation cultures in a series of 37 qualitative interviews with prominent supporters and critics of Louisiana's Master Plan, completed between 2016 and 2018 as part of an ongoing study of coastal adaptation in this region. The interviews furnished insight into the political history of the Master Plan, as well as revealing themes and knowledge constructions that figure prominently in the discourse of these separate communities. Although climate cultures have been studied previously using survey (Heimann 2019) and media analysis (Christmann et al. 2014) techniques, interviews are also methodologically well-suited for research about this phenomenon. In an interview setting, as participants are asked to reflect upon, and explain their support for or opposition to, a particular adaptation approach, they mobilize socially shared "vocabularies of motive" to account for the positions that they take (Vaisey 2009; Mills 1940). Such responses illuminate elements of group knowledge and also, as communicative acts, reproduce these shared interpretations.

I identified an initial group of participants through an analysis of local news coverage of coastal adaptation and the Master Plan. I contacted individuals who were frequently quoted as authoritative sources in news stories and then, through a "snowball" process, enlarged the pool of participants. The vast majority of these participants live and work in southeastern Louisiana, although I also interviewed two representatives of national environmental organizations and members of Louisiana's Congressional delegation (one currently in office and the other now working as a consultant) who spend substantial amounts of time outside the region. Most of the interviews lasted between one and two hours and, following transcription, I used a grounded analytic approach to identify core themes and points of differentiation between the two groups (Charmaz 2001). In addition to interviewing both supporters and critics, my sample included individuals from various institutional fields, including the economy (business leaders), politics (state administrators and elected officials), science (university researchers), and civil society (environmental and community advocates). To protect their anonymity, I refer to interviewees by occupation and position relative to the Master Plan (i.e. a "critical scientist"), rather than by name.

It is important to note that my targeted interview strategy is limited in certain ways. In the first place, the sentiments voiced in the interviews cannot be taken to represent patterns of public opinion about the Master Plan, more broadly. Such an analysis is beyond the scope of this chapter. Instead, I focus on individuals who engage directly with the planning process, either as supporters or as critics. Second, the interviewees are largely drawn from among the community's elite. They are well-educated, materially comfortable, and racially privileged, and therefore not demographically representative of the general population in a region characterized by high rates of poverty and social exclusion. The elite bias of this group reflects the planning process itself, which usually occurs in specialized settings that require extensive background knowledge for effective participation (although opposition has also developed among the primarily white, working-class residents of fishing communities along Louisiana's southeastern coast). In addition, members of Louisiana's vibrant environmental justice and equity community, who often have closer ties to marginalized groups and communities, tend to focus their energies on anti-toxics campaigning rather than on climate adaptation.

As Table 8.1 shows, supporters and critics of Louisiana's adaptation approach were not evenly distributed among different institutional fields.

Table 8.1 Interview sample by institutional field and position toward the Master Plan

	Supporters	*Critics*	**Total**
Economy	5	1	6
Politics	7	0	7
Science	3	7	10
Civil Society	10	4	14
Total	**25**	**12**	**37**

Source: Own representation.

Supporters tended to be local business leaders, state policymakers, and personnel at large environmental organizations (included in figures for civil society). Critics were overrepresented among university scientists and also included members of smaller and more progressive environmental and social justice organizations. This pattern is hardly neutral – from the perspective of political economy, it highlights inequalities in the political influence of supporters and critics of the Master Plan. Supporters tend to be rich in material resources and political connections; they are "repeat players" in decisions about land use and regional economic development (Logan and Molotch 1987). In addition, the absence of critics in the political field highlights the fact that the Master Plan has enjoyed strong bipartisan support in the Louisiana legislature (indeed, it received unanimous votes of approval in 2007 and 2012) and that it has been designated a priority by state government. The community of critics is thus comprised of more politically marginal figures, who generally do not "sit at the table" when decisions are made by state government.

Competing adaptation cultures

I now turn to the interviews themselves and describe points of difference between understandings of climate risk and resilience articulated by supporters and critics of the Master Plan. Neither group questions the imperative of climate adaptation, but they diverge in their evaluations of Louisiana's adaptation approach. Importantly, the discourse of these groups is less technical than it is political, meaning that supporters and critics of the Master Plan anchor their evaluations in broader conceptualizations of political economic relationships and priorities in the region. I unpack these evaluations by considering three themes: the relationship between structural protection and relocation, the role of the oil and gas industry in coastal restoration, and the character of the scientific research supporting the Master Plan.

Structural protection and relocation

One line of cultural differentiation between these groups runs through their perceptions of the community protection components of the Master Plan. As discussed above, the Master Plan outlines an adaptation strategy that combines wetlands restoration activities with the construction of new flood control structures, such as levees and floodwalls. Supporters of this approach emphasize the complementarity of these approaches. For instance, the *regional director* of a well-known environmental NGO, which works closely with the state of Louisiana to promote the Master Plan, explained in reference to the state's Coastal Protection and Restoration Authority:

> I think the formation of CPRA and the merging of levees with coastal restoration is hugely successful and important because in both cases,

`you're moving mud, you're moving dirt, you're changing hydrology. You're affecting the same landscape. You're affecting the same communities. These things actually do work in tandem.

The arguments in favor of this approach are not only technical in nature. Supporters of the Master Plan also emphasize that these restoration and protection components present entrepreneurial opportunities to local engineering firms and environmental design businesses. In this sense, climate adaptation is seen as a potential source of jobs and economic growth. This is particularly important given Louisiana's historical dependence on the energy industry as a source of revenue and employment. Jobs in oil and gas extraction and processing are still enormously important to the state, but the largest companies have shifted many management operations further west to cities on the Texas Gulf Coast. At the same time, much oil and gas extraction has moved offshore into the Gulf of Mexico as onshore deposits have been exhausted. In this context, a *state employee* charged with implementing Master Plan projects noted that "there was a recent statistic out there that said that water resource jobs in this state, particularly along the coastal areas, will be greater than oil and gas jobs".

The *director of a business development organization* in New Orleans expanded this point:

> for a period of time, and for us, actually, it will be decades, we will be doing essentially these large-scale construction projects. These projects will employ people, allow for opportunities if we do this right … [There is] the opportunity to create a new industry that essentially creates permanent jobs and revenues because of the potential to export technology and services that we're developing here.

In contrast, critics tend to be more skeptical of both the effectiveness and the economic benefits of structural protection. On the one hand, they question the wisdom of investing in permanent structures, given the fact that under certain future scenarios, climate-related sea-level rise may make these structures obsolete. Speaking of a proposed levee project near the city of Houma, Louisiana, one *environmental activist* explained:

> [The] Morganza to the Gulf [levee] is another part of the Master Plan. It's ten billion dollars. It stinks. It's just this big levee out in the middle of the water. (…) That's not going to work. That sucks for everybody in Houma. That's why we push hard on relocation, because we're going to need it for Houma because that levee ain't going to work.

A *researcher* who had worked with the state for a period of time before developing a more critical orientation toward the Master Plan reiterated this point in somewhat more technical language:

I think we could spend a whole lot less money if we found ways to help people migrate slowly inland rather than spending billions of dollars building structures that will be outdated by the time they're finished. Every inch of sea level rise makes every fixed height levee less safe. By the time you finish this extensive system, it will need serious improvements and elevations.

This critique highlights an alternative approach to adaptation articulated by many of the Master Plan's critics. This approach emphasizes nonstructural protection measures, ranging from the elevation of buildings to planned relocation to land use regulations that restrict development in flood-prone areas, as an alternative to massive investment in structural protections. Although these non-structural measures are discussed in the Master Plan, challengers fear that they are not given the same priority as structural measures. As *one social scientist* put it:

the Master Plan declares that they have a non-structural element. It declares a large amount of money. It appears that the large amount of money they expected would come as a result of hurricane declarations and, besides a few pages, there is just not a commitment to that.

Another social scientist presented this concern more bluntly:

I think there's an unspoken assumption that hurricanes will dislodge people and they'll have to move on their own nickel inland. Why should we spend money helping people and communities maintain their cultural and historical integrity, their religious networks, their family networks? We're just hoping that there will be some dramatic event that dislodges people rather than a process that's sane and safe and equitable.

The statements of these supporters and critics of the Master Plan suggest that disagreements about community protection and relocation strategies are shaped by differences in interests and values between these two groups, both of which are connected to the social and economic positions of group members. Supporters in the state's business community and in government place a high value on the potential economic opportunities created by structural protection projects, not least because they are in a position to benefit directly or indirectly by the revenue such projects might create. Critics, on the other hand, advocate for relocation over structural protection on the grounds of social equity, arguing that a focus on economic growth pushes the burden of adapting climate risks onto communities that are already highly vulnerable.

The role of the oil and gas industry

A second point of divergence between these adaptation cultures has to do with the role and responsibility of oil and gas companies operating in

Louisiana in relation to the development and implementation of the Master Plan. As pillars of the state's economy, major oil and gas companies have provided financial support to nongovernmental organizations that advocate for the Master Plan. As the *executive director of one NGO* explained,

> [companies] have been pretty supportive of what we are doing. They have people that work here. They have billions of dollars in infrastructure. Obviously, they have a vested interest in what happens here (...). There is no separation.

The oil and gas industry provides indirect support to the Master Plan, as well, through the GOMESA revenue sharing law described earlier in this chapter. However, it is also important to bear in mind that oil and gas exploration has increased the risks that this region faces. As noted, oil canals and pipelines have caused a significant amount of wetlands loss over time, undermining the "natural defenses" that have protected coastal communities from hurricane-related storm surges (Freudenburg et al. 2011). In Louisiana, some coastal parishes and levee boards have filed lawsuits against oil and gas companies, arguing that companies are liable for risks and damages that result from their activities.

These lawsuits represent a dividing line between adaptation cultures in the region. In my interviews, critics tended to agree that these lawsuits were both justified and productive, while supporters were far more skeptical and questioned the lawsuits on both practical and ideological grounds. One *engineer* who supported the Master Plan remarked that:

> Most people agree that the industry did cause damage and has contributed to wetland loss in the state. Even the oil industry people will pretty much admit that. But it's a whole other question whether they are still legally liable or responsible for that, because most of that work was done with permits or with landowner contracts that allowed them to do the work. My opinion is that, probably, they are not liable.

More vociferously, an *NGO board member* asserted:

> I'm not one who thinks the oil and gas industry owes us anything for their actions in the past. ... I think it's counterproductive. I don't want to be blamed for slavery. I had nothing to do with slavery. The same thing, the companies today had nothing to do with what happened thirty years ago.

Rather than emphasizing industry liability, Master Plan advocates tended to describe oil and gas companies as potentially important partners in restoration efforts. Noting that these businesses also faced risks of climate-related economic losses, one advocate noted, "if you think you're losing your land, you want to be part of a solution (...). So why be punitive if someone is being proactive?" They emphasized the need to bring the industry to the negotiating table and to identify points of common interest. Lawsuits, one elected *state representative* opined, were "selfish" and short-sighted because they could have "substantial [negative] effects on the state's ability to negotiate or work out a compromise with the oil companies to address this problem".

While members of the dominant coalition portrayed this consensus-oriented approach as reasonable and pragmatic, challengers offered a very different interpretation. They argued that oil and gas companies were using donations to NGOs and political influence with the state to protect themselves from pressures to admit responsibility for the environmental burdens of their activities. Industry donations, in the words of one coastal scientist, had transformed many environmental advocacy groups into "lapdogs (...) seeking oil company finance, funding, that sort of thing". They suggested that these activities were an extension of a long and well-documented history of political influence-seeking in Louisiana (Houck 2015). One *critic of the Master Plan* process articulated this concern at length:

> I think the industry needs to be held accountable. ... There's evidence that the industry knows that they were causing great harm but they chose not to make an issue of it and instead use the Legislature as an arm of the industry. They have a [public relations] wing ... [and] what they do is they run commercials during the Super Bowl about the importance of America's wetlands to the rest of the nation and how they're disappearing and how they need to be fixed, but they never admit responsibility for it and they never say that they're going to pay. If they admit responsibility, and that happens here, it will likely spread throughout the world.

These divergent visions of the industry's role and motivations also extended to the financing systems that had been put into place to support the Master Plan. There is a none-too-subtle irony in financing a climate adaptation program with revenues from offshore fossil fuel extraction, as the GOMESA law does. Although elected representatives and business leaders from Louisiana have been prominent voices in the community of climate sceptics in the United States, I did not encounter explicit denial of connections between fossil fuel use and climate change in my interviews. However, members of the dominant coalition tended to gloss over the contradictions of this financing system, emphasizing instead the urgency of adaptation needs and the pragmatism of the Master Plan approach. As one of my interviewees put it, "and the alternative is what? You forget it and let it go? You have to do something and nobody said it was going to be perfect".

In contrast, challengers were more critical of the financing approach that the state had developed, although they, too, struggled to articulate a plausible alternative approach. In fact, the contradictions embedded in this approach were a source of near-anguish to some of these critics, as illustrated by the statement of *one environmentalist* who was deeply involved in international climate politics:

> The GOMESA program is supposed to give Louisiana a greater share in offshore royalty revenues. ... So one might think, well great, the state's going to get money. That's the most popular belief. But what's the cost of that? It means that we have to extract more to get more and it's like a self-defeating prophecy. You have the eighth largest source of carbon that is up for auction and if that carbon gets released into the atmosphere, then it's going to harm the climate a lot more than the critical situation we're already in. ... This whole notion that we should extract more oil in the Gulf and get more money to save ourselves is an injustice because it also harms the rest of the planet. It's very self-serving.

Supporters and critics of the Master Plan were thus divided by very different perceptions of the role of the oil and gas industry. Supporters viewed the industry as a reasonable partner and favored approaches that aimed to identify common interests. In contrast, critics treated the industry with deep distrust and argued that the state's dependency on oil and gas revenues increased the climate-related risks that it and other regions face. Again, these different perceptions should be understood in the context of the position of these different groups. Supporters, including those quoted in this section, often worked directly with representatives from the industry, building relationships that created a foundation of trust and mutually beneficial exchange. Critics largely did not have these relationships, which made it easier for them to perceive the industry as an impersonal and corrupt force.

The character of scientific research

Finally, champions of the Master Plan frequently emphasize the quality of the scientific analysis and modeling that guides the state's selection of adaptation projects for funding. For example, several of my interviewees highlighted the state's investment in creating a scientific and technical organization, called The Water Institute of the Gulf (TWIG), and in hiring well-credentialed scientists and engineers to develop the ecological and hydrological models that inform the Master Plan. Some members of the dominant coalition even accounted for criticism of the Master Plan by suggesting that challengers were scientifically ignorant, as in the case of *one NGO leaders* who commented that:

> (...) some people don't even believe in the science. ... They have anecdotal science or they just say, 'No, I don't think that's what really happened.' You have to get them to believe that the science is there.

However, this interpretation oversimplifies critiques made by challengers to the Master Plan. Regardless of positioning, my interviewees emphasized the importance of scientific research in coastal adaptation planning. Challengers noted, though, that the research taking place at TWIG heavily favored the physical sciences and engineering, while largely excluding the social sciences from adaptation planning. They argued that this imbalance restricted the range of adaptation strategies that the state was able to consider and reinforced the business-oriented approach described above. One *critic*, herself a university-employed social scientist, explained that:

> Physical scientists and companies benefit when the science requires that there be dirt and mud moved, so those two forces of studying the coast physically (...) and the economic interest in solving the problem through activities that are able to bear profit for those companies has kept that aspect of the solution front and center.

Challengers also raised concerns about the independence of the state's scientific modeling team. One interviewee, also a *social scientist*, who worked for TWIG for a period of time, commented:

> Part of the reason for the creation of the Water Institute was to kind of make sure there was a little tighter control on the research than can be exerted over independent university scientists and social scientists. ... The state has taken steps to minimize the role of independent scientists in this process over the years.

These critiques show that divergence between these adaptation cultures has less to with the valued placed on science itself, and but rather hinges on perceptions of whether the activities of scientific research and modeling have been organized in appropriate ways. While supporters see the highly focused character of the scientific activities carried out by TWIG, and the relative consensus that this style of organization has produced, as advantages, critics raise concerns about the potential for diminished scientific autonomy and the absence of critical debate that would result from a multidisciplinary approach to adaptation (Turner 2009).

Discussion and conclusion

Group knowledge about climate risk and resilience is socially constructed, not objectively given, and varies across social and geographic space (Christmann and Ibert 2012). The climate cultures literature has analyzed how groups create shared understandings of climate change through discursive exchanges, generating communities of knowledge that exist in culturally proximate or distant relationships to one another (Heimann 2019). In this chapter, I have combined the climate cultures approach with a political economy framework to explore how the formation of discursive communities is also linked to place-based struggles in which groups with varying degrees of influence strive to advance their interests and goals in the process of climate adaptation. This synthesis is important because politics and culture, as much as we may wish to separate them analytically, blend together in practice (Adger et al. 2013). In Louisiana, the lines that divide supporters and critics of the state's strategy for climate adaptation are simultaneously cultural and political, having to do with values and interpretations as well as interests and influence, and the difference between these elements is not easy to tease out. For instance, business and political leaders who support Louisiana's Master Plan are attracted by a vision of adaptation that emphasizes economic growth and collaborative relationships between the state and industry – two goals that the political economic literature has identified as central motivators of business advocacy in cities and regions across the United States (Logan and Molotch 1987).However, critics of the Master Plan have brought concerns about social equity, business accountability, and scientific autonomy to bear – all issues that stand in tension to the goals of the Plan's supporters.

One important extension of this line of thinking has to do with how political economic struggles may heighten the apparent differences between climate cultures. As individuals compete for influence and attention in political settings, they often find it advantageous to emphasize these differences in stark and invidious terms. This process of "boundary work" may lead people to pull away from potential common ground and to use stereotypes and simplifications to characterize one another's positions (Lamont and Molnár 2002; Binder 1999). Hints of this process appear in the qualitative data that I have presented, such as in one critic's characterization of environmental groups that support the Master Plan as industry "lapdogs" or in supporters' derision of lawsuits against oil companies as "selfish". At the same time, social conflicts encourage people to "pick sides", increasing the social distance between groups and making it less likely that people will seek compromise or synthesis between rival cultural interpretations. Attention to boundary work has the potential to advance the notion of "culture as a relational space" by revealing how social relationships between groups may affect cultural relationships constituted in discourse (Heimann 2019).

Climate adaptation in coastal areas is particularly complex, and Louisiana is ahead of many other areas of the United States in preparing to manage the challenges associated with sea level rise and changing weather patterns (Moser et al. 2012). Although much of the scholarly work about adaptation in Louisiana has come from the natural sciences and engineering, adaptation is also a cultural and a political process. The empirical information that I have presented in this chapter aims to flesh out those latter dimensions. Understanding cultural and political divergence in Louisiana provides insight not only into this region but also contributes to a broader portrait of variations in climate adaptation cultures in the United States and around the world.

References

Adger, W. Neil, Jon Barnett, F. S. Chapin III, and Heidi Ellemor. 2011. "This Must Be the Place: Underrepresentation of Identity and Meaning in Climate Change Decision-Making". *Global Environmental Politics* 11 (2): 1–25.

Adger, W. Neil, Jon Barnett, Katrina Brown, Nadine Marshall, and Karen O'Brien. 2013. "Cultural Dimensions of Climate Change Impacts and Adaptation". *Nature Climate Change* 3 (2): 112–117.

Adger, W. Neil, Jouni Paavola, and Saleemul Huq. 2006. "Toward Justice in Adaptation to Climate Change". In *Fairness in Adaptation to Climate Change*, edited by W. N. Adger, J. Paavola, S. Huq, and M. J. Mace, 1–19. Cambridge, MA: MIT Press.

Berger, Peter L., and Thomas Luckmann. 1967. *The Social Construction of Reality*. New York: Anchor Books.

Binder, Amy J. 1999. "Friend and Foe: Boundary Work and Collective Identity in the Afrocentric and Multicultural Curriculum Movements in American Public Education". In *The Cultural Territories of Race: Black and White Boundaries*, edited by M. Lamont, 221–248. Chicago, IL: University of Chicago Press.

Callison, Candis. 2014. *How Climate Change Comes to Matter: The Communal Life of Facts*. Durham, NC: Duke University Press.

Carmin, Jo Ann, Kathleen Tierney, Eric Chu, Lori M. Hunter, J. Timmons Roberts, and Linda Shi. 2015. "Adaptation to Climate Change". In *Climate Change and Society: Sociological Perspectives*, edited by Riley E. Dunlap, and Robert J. Brulle, 164–198. New York: Oxford University Press.

Charmaz, Kathy. 2001. "Grounded Theory". In *Contemporary Field Research*, edited by R. M. Emerson, 335–352. Long Grove, IL: Waveland.

Christmann, Gabriela, Karsten Balgar, and Nicole Mahlkow. 2014. "Local Constructions of Vulnerability and Resilience in the Context of Climate Change: A Comparison of Lübeck and Rostock". *Social Sciences* 3 (1): 142–159.

Christmann, Gabriela, and Oliver Ibert. 2012. "Vulnerability and Resilience in a Socio-Spatial Perspective: A Social-Scientific Approach". *Raumforschung und Raumordnung* 70 (4): 259–272.

Coastal Protection and Restoration Authority of Louisiana. 2017. "Louisiana's Comprehensive Master Plan for a Sustainable Coast". Baton Rouge: Coastal Protection and Restoration Authority of Louisiana.

Couvillion, Brady, Holly Beck, Donald Schoolmaster, and Michelle Fischer. 2017. "Land Area Change in Coastal Louisiana 1932 to 2016: US Geological Survey Scientific Investigations Map 3381". Reston, VA: US Geological Survey.

Fligstein, Neil, and Doug McAdam. 2012. *A Theory of Fields*. New York: Oxford University Press.

Freudenburg, William R., and Robert Gramling. 2012. *Blowout in the Gulf: The Bp Oil Spill Disaster and the Future of Energy in America*. Cambridge, MA: MIT Press.

Freudenburg, William R., Robert Gramling, Shirley Laska, and Kai T. Erikson. 2011. *Catastrophe in the Making: The Engineering of Katrina and the Disasters of Tomorrow*. Washington, DC: Island Press.

Gieryn, Thomas F. 2000. "A Space for Place in Sociology". *Annual Review of Sociology* 26: 463–496.

González, Juan L., and Torbjörn E. Törnqvist. 2006. "Coastal Louisiana in Crisis: Subsidence or Sea Level Rise". *Eos: Transactions of the American Geophysical Union* 87 (45): 493–498.

Gotham, Kevin Fox. 2016a. "Antinomies of Risk Reduction: Climate Change and the Contradictions of Coastal Restoration". *Environmental Sociology* 2 (2): 208–219.

Gotham, Kevin Fox. 2016b. "Coastal Restoration as Contested Terrain: Climate Change and the Political Economy of Risk Reduction in Louisiana". *Sociological Forum* 31 (S1): 787–806.

Haedicke, Michael A. 2017. "Institutionalizing Coastal Restoration in Louisiana after Hurricanes Katrina and Rita: The Importance of Advocacy Coalitions and Claims-Making in Post-Disaster Policy Innovation". *Case Studies in the Environment.* https://doi.org/10.1525/cse.2017.000422

Heimann, Thorston. 2019. *Culture, Space, and Climate Change: Vulnerability and Resilience in European Coastal Areas.* New York: Routledge.

Heimann, Thorsten, and Bishawjit Mallick. 2016. "Understanding Climate Adaptation Cultures in Global Context: Proposal for an Explanatory Framework". *Climate* 4 (4): 59–71.

Hochschild, Arlie R. 2016. *Strangers in Their Own Land.* New York: New Press.

Houck, Oliver A. 2015. "The Reckoning: Oil and Gas Development in the Louisiana Coastal Zone". *Tulane Environmental Law Journal* 28: 185–296.

Lamont, Michèle, and Virág Molnár. 2002. "The Study of Boundaries in the Social Sciences". *Annual Review of Sociology* 28: 167–196.

Logan, John R., and Harvey Molotch. 1987. *Urban Fortunes: The Political Economy of Place.* Berkeley: University of California Press.

Lopez, John A. 2009. "The Multiple Lines of Defense Strategy to Sustain Coastal Louisiana". *Journal of Coastal Research* 54: 186–197.

Marshall, Bob, Al Shaw, and Brian Jacobs. 2014. "Louisiana's Moon Shot". *The Lens,* 8 December. http://projects.propublica.org/larestoration

Mills, C. Wright. 1940. "Situated Actions and Vocabularies of Motive". *American Sociological Review* 5 (6): 904–913.

Moser, Susanne C., S. Jeffress Williams, and Donald F. Doesch. 2012. "Wicked Challenges at Land's End: Managing Coastal Vulnerability under Climate Change". *Annual Review of Environment and Resources* 37: 51–78.

O'Brien, Karen L., and Johanna Wolf. 2010. "A Values-Based Approach to Vulnerability and Adaptation to Climate Change". *WIREs Climate Change* 1 (2): 232–242.

Paola, Chris, Robert R. Twilley, Douglas A. Edmonds, Wonsuck Kim, David Mohrig, Gary Parker, Enrica Viparelli, and Vaughan R. Voller. 2011. "Natural Processes in Delta Restoration: Application to the Mississippi Delta". *Annual Review of Marine Science* 3: 67–91.

Rudel, Thomas K., J. Timmons Roberts, and Jo Ann Carmin. 2011. "Political Economy of the Environment". *Annual Review of Sociology* 37: 221–238.

Sovacool, Benjamin K., Björn-Ola Linnér, and Michael E. Goodsite. 2015. "The Political Economy of Climate Adaptation". *Nature Climate Change* 5 (7): 616–618.

Theriot, Jason P. 2014. *American Energy, Imperiled Coast: Oil and Gas Development in Louisiana's Wetlands.* Baton Rouge: Louisiana State University Press.

Tierney, Kathleen. 2014. *The Social Roots of Risk: Producing Disaster, Promoting Resilience.* Stanford, CA: Stanford Business Books.

Turner, R. Eugene. 2009. "Doubt and the Values of an Ignorance-Based World View for Restoration: Coastal Louisiana Wetlands". *Estuaries and Coasts* 32 (6): 1054–1068.

US Global Change Research Program. 2018. "Impacts, Risks, and Adaptation in the United States: Fourth National Climate Assessment, Volume II: Report in Brief". Washington, DC: US Global Change Research Program.

Vaisey, Stephen. 2009. "Motivation and Justification: A Dual-Process Model of Culture in Action". *American Journal of Sociology* 114 (6): 1675–1715.

Yuill, Brendan, Dawn Lavoie, and Denise J. Reed. 2009. "Understanding Subsidence Processes in Coastal Louisiana". *Journal of Coastal Research* 54: 23–36.

9 Playing Hide and Seek

Adapting Climate Cultures in Troubled Political Waters in Georgia, United States

Julia Teebken

Introduction

What kind of strategies might actors develop to enforce their political aims when they are confronted with adversarial climate-cultural environments? Against the background of widespread and politically motivated climate skepticism, this chapter aims to make sense of informal political processes by asking how local policy practitioners deal with these environments in the state of Georgia. It explores different coping strategies and (hidden) channels through which planning and/or policies are being initiated. Coping strategies present a form of climate culture in terms of adjusted local policy planning that is context-specific to the Southeastern United States but may be transferred to other regions across the country.

The chapter builds on Heimann and Mallick (2016) explanatory framework and Fineman's (2010) concept of vulnerable political institutions. In this chapter, vulnerable political institutions refer to the limited adaptive capacity of state institutions to propose effective solutions and actively deal with complex problems, such as climate change, which can be due to different reasons (e.g. the overtly complex nature of a problem, limited resources or political competences, strong lobby interests).

The next section provides the theoretical background in which the study grounded. It is followed by a section on the methodology used to address the above research questions. The subsequent section then briefly provides some background information about the state of Georgia and its socio-ecological contextual conditions, knowledge environment as well as political-institutional context against which the results are to be understood. The main part of the chapter, which presents the results of the study, explains the coping strategies that local politicians have developed. Because of Georgia's adversarial political environment for climate governance, most debates have focused on the persistent nature of politically motivated climate skepticism in order to describe the supposed gap of climate policy in the Southeastern region of the country. Inaction is often descriptively explained through the bi-partisan political divide, which results in policy and institutional constraints, such as the lack of political will, funding and federal guidance.

DOI: 10.4324/9781003307006-14

Yet, climate policy efforts have manifested throughout the state. The penultimate section discusses climate policy in Georgia as an example of lived, contextual experience before the conclusions summarize the findings and provide a final interpretation.

Theoretical background: climate cultures from a perspective of vulnerable political institutions

Hulme (2009) argues that adapting to and mitigating climate change is ultimately a cultural question due to different "vantage points", or differences of perspective and contrasting interpretations of climate change. Different meanings are not just prevalent in different scientific narratives but have deeper origins that are grounded in various attitudes, ethical, ideological and political beliefs, as well as aspirations. Discordant voices about climate change have become obvious in different interpretations of the past, (the present) and competing visions of the future (Hulme 2009). These different interpretations may result in diverse courses of action. Hulme predicted that there is a risk of disconnecting climate from its cultural forms by framing it as overtly physical and global, thereby allowing "the idea of climate change to acquire a near infinite plasticity" (Hulme 2009, 28). Becoming aware of the cultural anchors and social meanings of the way "climate" is constructed is helpful for understanding the way collective decisions are made.

The explanatory framework outlined by Heimann and Mallick (2016) and the sociology of knowledge perspective by Heimann (2019) build upon this understanding of culture as decisive determinant through which vulnerability constructions and resilience practices can be studied and understood. In this line of reasoning, culture is an aspect of collectively shared knowledge, which at the same time is socially constructed and "always bound to the reduction of complexity" (Heimann 2019, 19). Viewing knowledge as aspectual and shared construction of actors and adopting a relational approach is not just helpful in explaining climate cultures in the global context but also useful in examining the differences in dealing with changing climate and political environments locally. Existing research has reflected upon the cultural dimensions of climate impacts and adaptation as a broader social phenomenon (e.g. Heimann 2019; Adger et al. 2013). Climate cultures describe different perspectives and cultural differences in dealing with climate change and are also reflected in different interpretations of vulnerability. This includes different perceptions of the potential threats as well as different coping strategies (for review of existing approaches within social sciences and the theoretical foundations of related concepts, see Heimann 2019, 9 ff.). The way local policy practitioners engage with climate change (and their political environment) is a distinctive form of climate culture that has not been reflected upon in detail (see Heimann 2019, 274).

As part of the analytical framework, this chapter takes a closer look at the (informal) institutional context of local climate governance. It focuses on how actors engage in political processes by developing and adjusting their practices. This is a distinctive form of climate culture in the political landscape that stands in interrelation with the climate-skeptic and/or conservative political environment. It can be seen somewhat in contrast to more conventional understandings prevalent in cultural studies, which perceive of climate cultures as social realities of different groups within society. This article suggests that these practices should be seen in the context of the vulnerable political institutions and limited adaptive capacity of the political system in Georgia to initiate change and effective responses to complex problems such as climate change. This perspective builds upon earlier works of vulnerability theorist Martha Fineman (2008). Political vulnerability refers to the exposure of governments and is characterized by interdependent institutions operating in a political economy that constantly redesigns vulnerability at various levels (also see Finan and Nelson 2009). According to this understanding, different systems of governance interact and governments themselves are vulnerable, for example in terms of being confronted by complex, unforeseen challenges, or "external conditions" over which the state has only limited or no influence. Examples are not limited to the sudden collapse of a bridge, corruption, or terrorist attacks, all of which can directly impact (the perception of) political priorities and may as a result lead to a shift in political agendas.

Political vulnerability is context-specific and determined by its political environment and socio-economic and physical context conditions. Both forms of vulnerability, human and political vulnerability, are ubiquitous and constant conditions of the human condition as well political systems (Fineman 2010). The article argues that these coping strategies can be understood as resilience practices in terms of vulnerable political institutions. Informing the research on climate adaptation cultures from a perspective of (informal) local decision-making and planning practices against the background of vulnerable political institutions helps to understand political processes in southeastern Georgia. It also advances the understanding about how governance is structured around issues of climate policy in adversarial political environments as a specific form of climate culture. In line with Kingdon (2011), this contribution adopts a broad understanding of policy practitioners as those who shape policymaking. This includes actors engaged in the designing of ideas, policy analysis, advice and planning and is not limited to elected government officials or state employees.

Methodology

This study builds upon data from a qualitative empirical study of 26 semi-structured expert interviews (Phase 1) and 5 problem-centered interviews (Phase 2) as well as participant observation. Experts act as "crystallization point", through inheriting practical insider knowledge which helps to reconstruct certain social situations or processes (Bogner et al. 2009). Problem-centered interviews aim to enable a structure for conversation that is process-oriented and directed toward specific problems (Witzel and Reiter 2012). Phase 1 interviews were held with elected government officials and governmental administrators (both considered "government officials"), policy planners and advisors from the non-state sector and academia, who either are involved or have expertise in (environmental) policy planning processes at different government levels in Georgia between October 2016 and May 2017 (for a full list of interviews, see Table 9.1).

Government officials working in related fields such as emergency management, city planning or public health and academics with expertise in climate policy were also interviewed. Participant observation and field notes from the attendance at the first Georgia Climate Conference (GCC) in November 2016 further grounded the study. The first conference brought together actors from across Georgia and other parts of the United States engaged in climate (policy) research and practice. The preliminary findings resulted in a refined interview guideline, which formed the baseline for a follow-up study. This follow-up, qualitative study was a targeted study and also consisted of participant observation (Brodkin 2017) during the second GCC in November 2019 and five problem-centered interviews thereafter. The audience of the second GCC was comprised of 430 experts from different levels of governance (public, private, non-profit) and academic sectors. The phase 2 interviews were conducted with local academics and policy practitioners involved in climate governance in Georgia.

Based on the number of non-scientific actors and different participants involved in the GCC such as those working for local municipal governments, Georgia entrepreneurs or political-administrative (state) government agencies, the conferences offered a rich ground for the study of a specific environment and ongoing political processes throughout Georgia.

The empirical data was systematically processed and clustered through open coding to detect main chunks of data, as prevalent in Grounded Theory (Strübing 2014).

Table 9.1 Interview samples by institutional field and primary level of political expertise

Expertise	County	Municipality	State	Across-country	**Total**
Government administration	2	4	2	0	**8**
Elected government	4	1	0	0	**5**
Civil society (Think tanks, NGOs)	0	3	0	7	**10**
Academia	0	7	1	0	**8**

Source: Own representation.

Study background: socio-ecological, knowledge and political-institutional context of Georgia

The analysis unveils different coping strategies of actors and participants of the Georgian Climate Conferences (2016 and 2019). The coping strategies enabled actors to push forward climate change related political action in Georgia despite dominance of adversarial and climate-skeptical cultural environments. Before describing the coping strategies, this section makes use of the culture-theoretical framework of Heimann and Mallick (2016) to heuristically sketch the socio-ecological, knowledge as well as political-institutional contexts of our actors' everyday socio-political lifeworlds.

Socio-ecological context. The Southeast United States is considered to be among the exceptionally susceptible regions to growing climate impacts such as sea-level rise, extreme heat events and precipitation, hurricanes and decreased water availability, as well as a long history of having to deal with climate-sensitive natural hazards (e.g. USGCRP 2018, 19 ff.; KC et al. 2015). Vulnerability assessments state Georgia as particularly vulnerable to climate extremes (KC et al. 2015). Droughts, extreme rain as well as storms have become more common, with some populations being more affected than others (KC et al. 2015).

Across the state of Georgia, the perceived pressure to act on climate change has been increasing. This has become especially visible in the context of water scarcity and ongoing political struggles over two shared river basins, the Apalachicola-Chattahoochee-Flint and Alabama-Coosa-Tallapoosa, between Alabama, Florida and Georgia. The legal conflict over the use of these basins began in the 1990s and has become known as the tri-state water wars. Both basins are critical for meeting metro Atlanta's water supply needs and have been confronted by increasing stress due to more severe and prolonged droughts. Unabated, these conflicts over, for instance, the use of resources, will pose more significant challenges in the future in light of intensifying climate change.

Knowledge context. Compared to other states, Georgia has historically had few climate policies and continues to have greenhouse gas emissions growth (Rabe 2016). The state's lack of a grand plan and lack of political leaders to address climate change is widely known and documented. Like other Republicans across the United States, most Republican decisionmakers in Georgia are described to be taking a "half pregnant position" characterized by not acutely denying that the climate is changing but questioning the human influence and taking a "drill baby drill" position (Galloway 2019; Chapman and Bluestein 2015).

Yet, there are notable regional variations of climate change beliefs, attitudes, perceptions and knowledge (Howe et al. 2015). Researchers argue that these variations correspond with a diversity of political environments for climate policy in terms of public climate change policy support and behavior. According to a study from 2020, 70% of the Georgian population

believe global warming is happening, which is only slightly below the national average of 72% (Marlon et al. 2020). However, only approximately
half (54%) of the Georgian population believe that global warming is caused
mostly by human activities and people have an even lower risk perception
with only 43% thinking that global warming will harm them personally
(Marlon et al. 2020). Although public opinion on the matter seems to be
divided relatively in the middle, policy support for a wide range of mitigation issues, such as regulating CO_2, is relatively high, especially when it
comes to renewable energy sources. Within political institutions, a shared
understanding about the impacts of climate change appears to be absent,
which impacts the way climate information is accessed, understood and
used (Bolson et al. 2013). Strong climate skepticism at the state level is a
specific knowledge context that has radiated into different political spheres
in Georgia and is differently pronounced at local-level jurisdictions.

Context of political institutions. Throughout the years, Georgia has become
more and more decentralized politically. This resulted in an increasing diver-
sion of responsibility to local jurisdictions and outsourcing of services from
county to city level. Decision-making is fragmented, as are jurisdictional
boundaries, which appear as core components of political decision-making
in Georgia and have impacted local actors to initiate change.

Against this institutional background, different policy efforts have un-
folded mainly at the local level. At the state level, few efforts exist, including
a greenhouse gas inventory initiated by the Georgia Department of Natural
Resources that also hosted the first GCC in 2016 and was responsible for in-
corporating sea-level rise (SLR) and shoreline/landslide stabilization in the
state-wide Hazard Mitigation Plan. A major recent development regards
the latest three-year plan (2019) on Georgia's energy mix put forward by the
Georgia Public Service Commission (PSC). As part of PSC's energy plan,
the share of solar power was increased to 2,210 megawatts of new solar,
which has been estimated to suffice for powering 200,000 homes (IEEFA
2019). This directed the monopoly Georgia Power to increase the share of
renewable energy (SELC 2020). Although this amount is relatively little,
compared to the number of PSC customers (2.6 million in 2019), it hints
to the increasing desire to change the energy mix of the state. The perception of decreasing economic viability of the current energy mix has further
pushed solar as core future utility in the Southeast. In addition to these
few efforts at the state level, some coastal counties and municipalities have
progressed in climate policymaking. Examples include the municipality of
Savannah and Chatham county, which have been working on a sea-level
sensors project that can be used for emergency planning purposes as well
as adaptation.

The coastal municipality Tybee island was the first to publish an adaptation plan in Georgia (2016). Apart from adaptation efforts at the coast,
municipalities and counties across Georgia have begun to include sustainability (and to a certain degree climate change) in their local planning. Aside

from popular examples such as Atlanta's Climate Action Plan from 2015, the Clean Energy Atlanta Plan from 2017 and Atlanta's Resilience Strategy from 2017, other inland located jurisdictions have committed to renewable energy targets (e.g. Oxford, Augusta, Clarkston). In 2019, renewable energy, including hydroelectric power, contributed nine percent of the state's utility scale generation (EIA 2020).

Results: coping strategies

As the previous section indicates and the majority of interviews emphasized, climate governance in Georgia has been signified by a general lack of political consensus at the state level, which for many interviewees was reflected in constantly low financial resources and the lack of political support exacerbated by limited federal support under the Trump administration. But even before Trump was elected, a speaker at the first GCCC (2016) highlighted the lack of political relevance: "the stars for climate change in Georgia are not perfectly aligned". Most work on climate change is informed by the politically divided nature of climate skepticism and the so-called "Georgian way" of dealing with things "from the bottom-up" (I-09, I-11). "Bottom-up" in the stated Georgian sense refers to low governmental interference from higher levels, highly decentralized policymaking signified by "multiple numbers of different governments" (I-22) and "not wanting to be told what to do" (I-11), which carries the connotation of a widespread non-conformist and state-skepticism culture.

The findings suggest that policy practitioners throughout Georgia managed to initiate climate policy developments despite these challenging contextual conditions. One overarching observation is that climate policy progress is unevenly distributed throughout the state and more visible at the coasts and in the main municipalities (I-11). Another observation is that much has happened since the first GCC in 2016. Here, the atmosphere seemed to be one of carefully building momentum as expressed by *one speaker at the conference*:

> Is Georgia Climate Ready? I can probably say no. But no local government, no state, no country is. We are not behind on this – we may not be on the forefront – but also not on the bottom of the list.

In contrast to 2016, the language changed with some conference participants actively pushing forward a narrative of change during the second GCCC in 2019, as expressed by *this speaker*:

> There are four pedestals through which Georgia is taking leadership: science, stronger conversation, solutions, and a stronger network. (...) I definitely think we are a leader on climate change, we could be a much more international leader.

The message of Georgia's (recent) leadership was iterated throughout the GCCC 2019 speeches as well as the qualitative interviews. It became evident that, between 2016 and 2019, actors throughout the state had managed to build and strengthen a network on the issue of climate change and to push forward climate change related political action. What made these actors successful in such a climate-skeptical environment? The findings indicate that the actors developed specific social and communicative strategies. In the following, some of these strategies are sketched in order to understand these (political) processes. It needs to be emphasized that this is a first outline, which should be deepened and justified by further research.

Social level: building creative coalitions and new networks

Throughout Georgia, different actors talked about the lack of support from higher-level government and emphasized their focus on local governments, private funders and foundations. The state government was reported as an unreliable source, why actors decided to not rely on them to take (political) action in Georgia (I-08, I-27). Against this background, local policy planners, advisors, NGOs and academics sought of different ways to engage on the issue of climate change by building strong actor coalitions, as this policy practitioner at the state level expresses:

> (...) these are our partners: Georgia Conservancy, Nature Conservancy, [...] We try to bringing together our resources rather than duplicate or divide NGOs – that would be a pain in the side. [...] We try to bring them to the table when we coalesce with other organizations such as the University of Georgia, Georgia Southern University or key local governments that are committed to be the champions at the coast: Tybee island, Campton county, Liberty county, Glynn county.
>
> (I-11)

Actively pursuing the issue of climate change by building and maintaining actor coalitions throughout Georgia resulted in "lifting of Georgia from being a blank spot" (I-21). As part of this strategy, actors have focused "on choosing the audience wisely" (I-27), e.g. "most commonly working with local governments, sustainability NGOs, mayors". One of the reasons stated had been that these "organizations are receptive to these ideas" (I-31). A popular example is the Georgia Climate Project (GCP), which is a state-wide consortium of people working on climate change in Georgia. The Fourth National Climate Assessment (2018, 756) referred to the "cross-disciplinary group" of the GCP as an example of best practice by developing research roadmaps that can help policy practitioners to prioritize research and action.

As part of the GCP, the Georgia Climate Series was launched, which highlights the personal stories of Georgians and how their livelihoods are impacted by climate change. "Drawdown Georgia" is another example set forth by the GCP, which aims to identify a set of solutions for the state of Georgia through the provision of a knowledge portal, the establishment of multi-disciplinary working groups and the selection of high-impact solutions. Another example is the Athens-based Georgia Climate Change Coalition (GCCC), which is comprised of environmental activists, community organizers, as well as students and academics from the University of Georgia. GCCC has been an advocate for creating a transition community on climate change.

Both projects are exemplary for the outstanding role educational actors fulfill in Georgia. University actors have not just taken on the classical role of providing science and actively informing policy efforts at different levels of government and the non-state sector but have also been key in reconvening and engaging local communities. This includes a new generation of changemakers and students, who for instance provided advise for local municipal actors on their climate action planning in Atlanta or running their own community projects.

Beyond that, Georgia's policy efforts on solar have relied on pressure exhibited by educational actors, their networks and unusual actor coalitions coupled with the engagement from a variety of different organizations. This includes State Republican Tim Echols, Republican State Senator Chuck Hufstedter, the Southern Environmental Law Center and faith-based, as well as community organization such as Georgia Interfaith Power and Light and Partnership for Southern Equity. Republican Tim Echols has actively pushed the solar energy agenda and expressed at the second GCC: "I know that we Republicans do get blamed for not using the c word. Do not paint with a broad brush".

Another unusual actor constellation includes the so-called "Green Tea Coalition", which is a bi-partisan coalition of the environmental NGO, the Sierra Club and the Tea Party, founded in 2013. The Coalition has been actively fighting for state engagement on solar power irrespective of their different political and ideological beliefs as well as organizational agendas. The Coalition managed to find common ground in challenging the monopoly of Georgia Power, the utility company in charge of electric utility.

Creative actor coalitions also resulted in overcoming some points of bi-partisan divide, resulting in an agreement on the adoption of electric cars and building larger state fleets (I-30), making Georgia become the leading market for electric vehicles. Although motivations might be different to pursue a renewable energy agenda, there appears to exist consent over underlying issues, which were then linked to solar energy.

Knowledge and communication level: language, narratives, issues

A great amount of policy advisors and those engaged in climate action appear to cope with their politically unfavorable environment through the use of three communication strategies: (i) their adjustment of language and use of non-climate frames, (ii) building narratives and audience-specific frames, and (iii) linking climate-related policy issues to local issues and cultural concerns. These key communication strategies also draw from the social level, and the way knowledge on this matter was shared as part of informal actor-networks.

(i)   Language adjustment and use of non-climate frames

The adjustment of everyday language has played a significant role throughout Georgia as this academic and policy advisor expressed:

> one thing, which has been a really difficult adaptation, I guess, is often avoiding the term climate change, which comes as internal friction because as a scientist you want to call it what it is [...]. But when trying to speak particularly to let's say a more conservative audience using the term creates a defense mechanism and brings a whole host of other issues up. So, to avoid getting into political debates and distractions, a lot of times, the message is framed differently. Maybe we talk about how agriculture is changing or the impact on forestry or national security, challenges and threats to the military and threat multipliers. So, all those terms are pulling to the same thing without having to get into the debate.
>
> (I-27)

Policy practitioners are partially engaged in practices of backdoorism, i.e. disguising climate change issues or policy interests as different topics due to their politically sensitive context. A non-state actor and local politician expresses:

> We tend not to start out the conversation as climate change, but growth first. Once they [Republicans and climate conservatives in charge of big organizations such as Georgia Power] realize the benefit, they open up their minds more. If they choose to not believe climate change that is not the problem, as long as they see the benefits. There was this rural community, Taylor county in rural Georgia (...), basically the old-timers, we never thought to get them on board. But now the county government is getting additional tax revenue [from solar], making believers out of people once they see economic benefits.
>
> (I-08)

Aside from local-level politicians, organizations, scientific advisors and NGOs, the preliminary findings suggest that it is also specific actors in the hierarchy of the political administration, who work for state agencies and governmental organizations and are liable to higher levels in the organization that make use of these coping strategies. One person working in a regional planning and intergovernmental coordination agency indicated that certain policy aspects are being budgeted

> (...) under different columns, in order to not try to cause too much at-tention from the subordinates (...) And when we went for the federal grant [...] we marketed it as extreme weather resilience as just a way of protecting the program.
>
> (I-28)

The avoidance of certain vocabulary and use of non-climate frames was widely talked about. The adjustment of language in daily operations in-cluded describing "how the environment is changing over long periods of time" instead of talking about climate change (I-20, I-28). One government employee in a state agency expresses:

> We do not go out and say: 'you terrible coal plant.' If we did that, there would be no progress. We gotta pick our approach wisely.
>
> (I-11)

Others implied that they had a list of cover words, such as resilience, that contain climate-sensitive vocabulary and potential substitutes (I-28, I-30). There also appeared to exist a form of exchange with culturally like-minded on language adjustments, as this person expresses:

> there are peers, like conversations we would have in certain workgroup meetings (...) that would be an opportunity to discuss how we are word-ing things or how we would be dealing with these kinds of issues, kind of copy each other a bit.
>
> (I-30)

The way language is adjusted also appears to depend on the perception of vulnerability categories. Awareness was strong about what were considered so-called "traditional vulnerabilities" such as poverty, drug addiction and vulnerability to crime (I-31). The pre-existing awareness about an estab-lished topic may correlate with greater openness toward putting the issue onto the agenda. The exacerbation of a traditional problem such as a dis-ease is likely going to receive more attention than topics such as heat or climate-induced diseases such as the Zika or West Nile Virus (I-22, I-29).

Language adjustments and the marketing of information is part of the everyday business of people working in politics and administrative organiza-tions. As part of their work in hierarchical organizations, pro-environmen-tal actors are bound by context factors which they adapt to by packaging the information differently. This act of translating knowledge in order to impact sense-making of decisionmakers is key to this communication strategy. One interviewee pointed out that backdoor smuggling of climate agendas through the adjustment of language and other (more traditional) policy topics was especially difficult to prove, as "you never know for sure" about the intention through which policies were initiated and or policy planning occurred (I-28).

(ii) Building narratives and audience-specific frames

Despite the adjustment of language in daily work routines, actors engage in a careful choice of audience-targeted frames and storytelling and have become increasingly creative in the way they narrate and deliver messages, as "stories beat statistics" (speaker at the GCC 2019). This includes presenting certain information using metaphors, comics and/or visuals to build a stronger narrative but also adjusting frames to the audience, building trust and through that strengthening ties toward distinctive audiences to be able exchanging over certain issues. This academic and policy practitioner working in agriculture expresses:

> Georgia is a pretty conservative state as I am sure you know. And the conservativeness of the politics also follows, to some extent, the conservativeness of the farmers. And so, I have to be careful what I talk about because I do not want to lose the audience. [Communication adjustment] depends on what kind of farmers I am talking to.
>
> (I-29)

Audience-targeted frames and storytelling is not an exclusive Georgia phenomenon. Actors throughout the state can fall back upon a great variety of existing resources and insights from the science of climate risk communication provided by the National Oceanic and Atmospheric Administration, or climate communication pieces provided by the Yale Climate Communication Center, as well as the platform Climate Communication and Outreach. Another academic who is also working on communicating the science and working with policymakers on solutions explains their use of different frames in Northern Georgia:

> I do not avoid it [using the term climate change] if I can. But when I am in front of a very religious group I would talk about it differently, by using Eco-Theology concepts. With a group of farmers, I would ask: do you know the weather is changing? So, more focused on the impacts. Another group is the retirees and older generation. Here, we make a conscious effort to speak to those groups. (...) The public administration and political actors are the toughest audience I have encountered.
>
> (I-27)

Others referred to geographical difference in targeted storytelling, with the Northern part of Georgia being distinctively more conservative, requiring a more careful adjustment of language. Furthermore, interviewees stressed that municipalities such as Atlanta are facing a different political reality than elsewhere because of its independence, which was also reflected in different patterns of communication (I-27, I-31). Nevertheless, this did not imply that people were entirely outspoken on climate change related matters, as one interviewee at the Mayor's Office of Resilience indicated (I-15).

(iii) Issue linking to culturally prevalent topics

A third communication strategy is the linking of climate policy topics to specific cultural issues and local themes. This strategy is characterized by the pairing of climate issues with conservative and traditionally Southeastern topics, such as property rights or the free-market economy. In the context of mitigation, solar advocates aren't just selling solar as a way to reduce emissions or reduce fossil fuels. Solar has been positioned as a property rights issue "pitting private citizens against utilities, regulators and fixed rates of return" (Kanellos 2013). The free-market narrative was successful in appealing to a long southeastern tradition and linking up with renewable energy.

In the context of climate adaptation, the linking of climate change (narratives) to locally specific farming backgrounds was talked about, such as minority, low-income and substitute farmers, who are interested in maintaining the family business also 20–50 years from now in contrast to big production farmers or livestock farmers, who are struggling with shorter time-frames (I-29).

Interviewees implied that there existed a list of political sensitive issues and that these were context- and work-environment specific. For instance, air quality and resilience and to some degree climate adaptation were considered less controversial issues than affordable housing or rising sea levels (I-11, I-28).

> It's all about the message (...). Free market, competition, choice, expanding the energy portfolio, and energy mix. I don't want excessive regulations.
> (Georgia Tea Party Activist Debbie Dooley, Inside Climate News)

The interviewees suggested that these issues are more appealing to a wider audience and a good fit to issue-link the climate agenda.

Discussion: climate policy as an instance of lived, contextual experience

The results show that different climate cultures play a vital role in climate governance throughout Georgia and demonstrate how climate-conscious policy practitioners developed an array of strategies to act in "hostile" climate-cultural contexts.

The preliminary findings suggest that climate-conscious people are to be found in a variety of different organizations such as federal agencies based in Georgia, intergovernmental regional organizations, NGOs and different academic institutions that have been advising policymakers. Climate-consciousness is also found in various regional contexts, signified by different degrees of climate skepticism and conservativism, e.g. the rural countryside or Georgia's north.

Adjusted language and narration are central elements of the coping strategies used by policy practitioners. Although actors have found ways to communicate climate science through other channels and thereby initiated action, the avoidance of climate-related terms and strategizing on these matters also points to the lack of a shared understanding and lack of intentional action. This is in line with literature that emphasizes different discourses as emblematic of divergent local knowledge resulting from different local experiences with economic problems and social marginalization (e.g. Christmann and Heimann 2017). In Georgia, local experiences with climate impacts are very different within society and among political actors. This experience is not just determined by local experiences with economic problems but shaped by uneven access to public goods such as health, education and transportation grown from a long path-dependency of racially motivated segregation and political marginalization of certain groups. Studying the different discourses related to (non-)climate policy-making may provide further insights not just on discourse cultures but also enhance our understanding of historic path-dependencies and how they impact climate cultures as part of political decision-making.

The interviewed policy practitioners appear to be better prepared through the strategies they developed and have grown more independent of state-based support or have disguised their policy efforts from within by using certain communication strategies. The findings however also suggest that the adjustment to climate denying environments has come at a considerable cost in terms of more lengthy political processes of initiation, little policy enforcement mechanisms and gaps in policy implementation as well as attention to only certain issues that satisfy an economically beneficial logic.

In Georgia, policy practitioners grapple with the institutional structures and dynamics of a highly decentralized political system with little guidance from the top. Going back to the notion of vulnerable political institutions, it has become obvious that the political system of the United States and that of Georgia, in particular, has relatively low institutional capacity to address the problem of climate change in a stable, predictable and concerted manner. The need for a change of organizational structures and routines becomes visible.

Despite the widespread acknowledgement that climate policy is a localized phenomenon in the United States, policy practitioners and some state actors in Georgia do perform an important role in climate governance that is underappreciated. Many local policy practitioners in Georgia appear to be internally driven by their political commitment to the cause, some even because of their adversarial environment. The look at alternative strategizing in terms of developing coping strategies might offer new insights into the role informal policymaking can and does already play for pushing climate action in a divided state.

Conclusion

This chapter examined informal political processes of local climate governance in Georgia. It offers some answers to the question what kind of coping strategies actors may develop, when they are confronted with deviant climate-cultural contextual settings. Local policy practitioners developed different coping strategies as a result of their contextual political environment, which can be considered an instance of adapted climate culture. Despite an unfavorable political-institutional context and knowledge environment when it comes to climate change (risk) perceptions and beliefs, these policy practitioners managed to push climate action forward anyhow because of the strategies they have developed throughout the years. Two distinctive coping strategies were detected: at the social level, creative actor constellations and unusual stakeholder alliances were built, with Georgia now consisting of a much broader network to enhance the climate agenda. This includes support from Republicans and private sector alliances and a strong network of educational actors that formed over the past six years. The second strategy relates to the knowledge and communication level. Here, three strategies stand out: (1) the adjustment of language, (2) audience-specific framing and narration, (3) as well as issue-linking to dominant (local) cultural issues.

The findings are preliminary and suggest that these strategies are differently pronounced geographically and across policy practitioners. Further, the strategies take different shapes in state agencies and policy sectors engaging in climate policy development. Thereby, the study challenges the prevailing image, which has constructed Georgia as a blank spot on the map of climate policymaking in the United States. These results are in line with the culture-relational understanding of Heimann (2019, 27) who suggests to think out of the box of absolute or container-spaces in order to retrieve a much more differentiated understanding of cultural spaces.

Research on climate policymaking in the Midwestern United States suggests that this region is characterized by similarly climate-skeptical and political conditions but conveys climate adaptation and mitigation strategies (e.g. Doll et al. 2017). Future research could analyze in greater detail, if the communication strategies differ geographically across Georgia and how effective they are in influencing the sense-making of decisionmakers and thereby impacting policymaking. Assuming, that it is not only in "climate-skeptic states" in the Midwest and Southeast where actors have developed coping strategies, the development of non-climate-frames would also be interesting to examine in states with stronger climate awareness such as California.

Finally, the paper contributes to the climate cultures debate by introducing research on coping strategies when it comes to the interaction between actors with different knowledge backgrounds. This deepens our understanding of local policy processes, some of which lie outside or overlap with the formalized channels of partially non-responsive public governance structures at the state level. Further research could examine what motivates policy practitioners to develop coping strategies across policy sectors.

References

Adger, Neil W., Jon Barnett, Katrina Brown, Nadine Marshall, and Karen O'Brien. 2013. "Cultural Dimensions of Climate Change Impacts and Adaptation". *Nature Climate Change* 3: 112–117.

Bogner, Alexander, Beate Littig, and Wolfgang Menz, eds. 2009. *Experteninterviews. Theorie, Methoden, Anwendungsfelder* [Expert Interviews. Theory, Methods, Fields of Application]. Wiesbaden (Germany): Verlag für Sozialwissenschaften.

Bolson, Jessica, Christopher J. Martinez, Norman E. Breuer, Puneet Srivastava, and Pamela Knox. 2013. "Climate Information Use among Southeast US Water Managers: Beyond Barriers and toward Opportunities". *Regional Environmental Change* 13: 141–151.

Brodkin, Evelyn. 2017. "The Ethnographic Turn in Political Science: Reflections on the State of the Art". *Political Science and Politicsm* 50 (1): 131–134.

Chapman, Dan, and Greg Bluestein. 2015. "A Rising Tide of Concern. State Agency's Warning on Climate Change in Georgia Spurs Action, Skepticism". *Atlanta Journal Constitution*, August 9, 2015. http://specials.myajc.com/rising-tide/

Christmann, Gabriela, and Thorsten Heimann. 2017. "Understanding Divergent Constructions of Vulnerability and Resilience: Climate-Change Discourses in the German Cities of Luebeck and Rostock". *International Journal of Mass Emergencies and Disasters* 35 (2): 120–143.

Doll, Julie E., Brian Petersen, and Claire Bode. 2017. "Skeptical But Adapting: What Midwestern Farmers Say about Climate Change". *Weather, Climate, and Society* 9 (4): 739–751.

Energy Information Administration (EIA). 2019. "Georgia State Profile and Energy Estimates". *EIA*, November 19, 2020. https://www.eia.gov/state/?sid=GA

Finan, Timothy J., and Donald R. Nelson. 2009. "Decentralized Planning and Climate Adaptation: Toward Transparent Governance". In *Adapting to Climate Change*, edited by Adger et al., 335–350. New York: Cambridge University Press.

Fineman, Martha A. 2008. "The Vulnerable Subject: Anchoring Equality in the Human Condition". *Yale Journal of Law & Feminism* 20 (1): 8–40.

Fineman, Martha A. 2010. "The Vulnerable Subject and the Responsive State". *Emory Law Journal* 60: 10–130.

Galloway, Jim. 2019. "Georgia Republicans Tiptoe toward the Conclusion that Climate Change Is Real". *Atlanta Journal Constitution*, April 9, 2019. https://www.ajc.com/blog/politics/georgia-republicans-tiptoe-toward-the-conclusion-that-climate-change-real/6eMae8zrTwfLd16ItQqMfN/

Heimann, Thorsten. 2019. *Culture, Space and Climate Change. Vulnerability and Resilience in European Coastal Areas*. London (United Kingdom): Routledge.

Heimann, Thorsten, and Mallick, B. 2016. "Understanding Climate-Adaptation Cultures in Global Context: Proposal for an Explanatory Framework". *Climate* 4 (4): 1–12.

Howe, Peter, Matto Mildenberger, Jennifer Marlon, and Anthony Leiserowitz. 2015. "Geographic Variation in Opinions on Climate Change at State and Local Scales in the USA". *Nature Climate Change* 5: 596–603.

Hulme, Mike. 2009. *Why We Disagree about Climate Change.* Cambridge (United Kingdom): Cambridge University Press.

Institute for Energy Economics and Financial Analysis (IEEFA). 2019. "State Regulators Tell Georgia Power to add 2,210MW of New Solar by 2024". *IEEFA*, July 17, 2019. https://ieefa.org/state-regulators-tell-georgia-power-to-add-2210mw-of-new-solar-by-2024/

Kanellos, Michael. 2013. "Behind the Tea Party Push for Solar in Georgia". *Forbes*, July 16, 2013. https://www.forbes.com/sites/michaelkanellos/2013/07/16/behind-the-tea-party-push-for-solar-in-georgia/

KC, Binita, Marshall Shepherd, and Cassandra Johnson Gaither. 2015. "Climate Change Vulnerability Assessment in Georgia". *Applied Geography* 62: 62–74.

Kingdon, John W. 2011. *Agendas, Alternatives and Public Policies.* Boston, MA: Longman.

Marlon, Jennifer, Peter Howe, Matto Mildenberger, Anthony Leiserowitz, and Xin-ran Wang. 2020. "Yale Climate Opinion Maps 2020". *Yale Program on Climate Change Communication*, September 2, 2020. https://climatecommunication.yale.edu/visualizations-data/ycom-us/

Rabe, Barry G. 2016. *Introduction: The Challenges of US Climate Governance.* Working Paper of the Brookings Institution.

Reiter, Bernd. 2013. "The Epistemology and Methodology of Exploratory Social Science Research: Crossing Popper with Marcuse". *Government and International Affairs Faculty Publications* 99.

Southern Environmental Law Center (SELC). 2020. "Georgia Public Service Commission Delivers Clean Energy Wins". *SELC*, July 16, 2019. https://www.southernenvironment.org/press-release/georgia-public-service-commission-delivers-clean-energy-wins/

Strübing, Jörg. 2014. *Grounded Theory. Zur sozialtheoretischen und epistemologischen Fundierung eines pragmatistischen Forschungsstils* [Grounded Theory. On the Social Theoretical and Epistemological Foundation of a Pragmatist Research Style]. Wiesbaden (Germany): Springer Fachmedien, VS Verlag für Sozialwissenschaften.

USGCRP. 2018. *Impacts, Risks, and Adaptation in the United States: Fourth National Climate Assessment, Volume II.* Washington, DC: US Global Change Research Program.

Witzel, Andreas, and Herwig Reiter. 2012. *The Problem-centred Interview.* London (United Kingdom): Sage.

Index

Printed in the United States
by Baker & Taylor Publisher Services

Printed in the United States
by Baker & Taylor Publisher Services